Organizational Culture and Absorptive Capacity

Dorothée Zerwas

Organizational Culture and Absorptive Capacity

The Meaning for SMEs

Dorothée Zerwas
Koblenz-Landau, Germany

Universität Koblenz-Landau, Dissertation, 2014
Vollständiger Abdruck der vom Fachbereich 4: Informatik der Universität Koblenz-Landau genehmigten Dissertation.

ISBN 978-3-658-05551-6   ISBN 978-3-658-05552-3 (eBook)
DOI 10.1007/978-3-658-05552-3

The Deutsche Nationalbibliothek lists this publication in the Deutsche Nationalbibliografie; detailed bibliographic data are available in the Internet at http://dnb.d-nb.de.

Library of Congress Control Number: 2014936104

Springer Gabler
© Springer Fachmedien Wiesbaden 2014
This work is subject to copyright. All rights are reserved by the Publisher, whether the whole or part of the material is concerned, specifically the rights of translation, reprinting, reuse of illustrations, recitation, broadcasting, reproduction on microfilms or in any other physical way, and transmission or information storage and retrieval, electronic adaptation, computer software, or by similar or dissimilar methodology now known or hereafter developed. Exempted from this legal reservation are brief excerpts in connection with reviews or scholarly analysis or material supplied specifically for the purpose of being entered and executed on a computer system, for exclusive use by the purchaser of the work. Duplication of this publication or parts thereof is permitted only under the provisions of the Copyright Law of the Publisher's location, in its current version, and permission for use must always be obtained from Springer. Permissions for use may be obtained through RightsLink at the Copyright Clearance Center. Violations are liable to prosecution under the respective Copyright Law.
The use of general descriptive names, registered names, trademarks, service marks, etc. in this publication does not imply, even in the absence of a specific statement, that such names are exempt from the relevant protective laws and regulations and therefore free for general use.
While the advice and information in this book are believed to be true and accurate at the date of publication, neither the authors nor the editors nor the publisher can accept any legal responsibility for any errors or omissions that may be made. The publisher makes no warranty, express or implied, with respect to the material contained herein.

Printed on acid-free paper

Springer Gabler is a brand of Springer DE.
Springer DE is part of Springer Science+Business Media.
www.springer-gabler.de

This thesis is dedicated to my parents Anke and Horst and my sister Claire for their endless love and encouragement.

Many thanks to …

… my advisor Prof. Dr. Harald F. O. von Kortzfleisch; through good and bad times, your kindness and comprehensive support have been ever-present in this important time of my life, for which I am deeply grateful.

… my co-advisor Dr. Klaus G. Troitzsch, university professor (retired); over the years you have helped me to find my way. You have pointed me in the right direction.

… my family and friends, for a life filled with laughter, happiness and fun times.

Foreword

In this book, Dorothée Zerwas highlights the need for small and medium-sized enterprises (SMEs) to build-up certain skills – their "absorptive capacity" – in order to be able to integrate strategically relevant knowledge that can only be found outside the organization. From the perspective of transfer, knowledge as a strategic resource also requires special treatment.

The organizational culture is a significant factor for successful transfer-related knowledge management and the consequent need to build up absorptive capacity in SMEs. The author has clearly indicated that there is, at this point, a great need for research in this area.

Many studies have focused on absorptive capacity. Zerwas has successfully managed to work out the manifold aspect- and theory-driven facets of the construct of absorptive capacity, in the context of SMEs. It is also clear that organizational culture has not yet played a significant role in the context of absorptive capacity.

A separate and important line of discussion has established the theme of organizational culture in the organization and management literature. In particular, Zerwas provides an excellent overview of the articles in the literature that have dealt with the importance of organizational culture for knowledge transfer and knowledge management. However, it is clear that a systematic, differentiated and comprehensive consideration of the effect ratio of organizational culture has not been made for absorptive capacity.

In connection with this, Zerwas has identified a multiple-founded research gap. First, SMEs are generally under-researched, but especially in the context of absorptive capacity, although the publication density on this issue is generally very high. Second, the particularly limited resources available to SMEs means that their absorptive capacity is generally weaker than it is for large enterprises, which is to be considered as a special feature. Finally, there is a thematic gap in the research literature in that the role of organizational culture for the absorptive capacity of SMEs has largely been neglected and has not been sufficiently differentiated.

As a result of the above-mentioned issue, the central research question in this book is as follows: How is organizational culture related to the acquisition capability, assimilation capability, transformation capability and exploitation capability of a SME and how should a knowledge-oriented organizational culture be designed to support the absorption of external knowledge? In order to answer this research question, the author describes in detail the two main concepts that are set in relation to each other: absorptive capacity and organizational culture. First, the foundations are laid for absorptive capacity, which can be mainly attributed to *Cohen & Levinthal* (1990) and *Zahra & George* (2002). In particular, the antecedent and components of absorptive capacity and the four special capabilities of the absorptive capacity (acquisition, assimilation, transformation and exploitation) are discussed in detail according to *Zahra & George* (2002). Second, the discussion of organizational culture stands in the foreground. The starting point is *Schein*'s (1984) concept of organizational culture. However, the publication density on the subject makes it clear that a far more differentiated consideration must be made as the representation of the basic model. To meet this requirement, Zerwas describes the six dimensions of a knowledge-friendly culture that *Sollberger* (2006) identified: trust, collaboration, openness, autonomy, learning receptivity, and care.

The focus then shifts towards the theoretical framework. Zerwas' most remarkable scientific achievement in this area is the initial identification of several organizational and management theories in the literature that contribute to the understanding of the essence of the concept of absorptive capacity. Following the theoretical foundation, the parameters of the model are worked out.

According to the central question of this research, each dimension of the organizational culture is then argued with regard to its impact on each of the capabilities of absorptive capacity.

The next chapter provides a comprehensive empirical analysis, followed by factor, correlation and regression analyses. The well-founded exclusion of structural equation models and the preferential treatment of the above-mentioned analyses emphasize the foreseeable high quality of the data collected. In the context of regression analysis, the central hypotheses of this research are reviewed, discussed, and finally turned into design recommendations for theory and management practice.

Dorothée Zerwas accepted the post of an assistant to my chair and lecturer in my research group two and a half years prior to publishing this book. The fact that she managed to do this in less than three years shows the intensity and perseverance with which she has pursued her doctoral studies while maintaining a high level of quality. It is particularly noteworthy that she was sustainably incorporated into the department administration, self-administration of the faculty, and teaching in my research group. Add to that the co- organization of the workshop "Entrepreneurship with a Migration Background" (approximately 100 participants) in the spring of 2012, and the primary organization of the largest entrepreneurship conference in Germany, the 17th Annual Interdisciplinary Entrepreneurship Conference (G-Forum 2013) in Koblenz, which had in excess of 300 participants. Dorothée Zerwas was also involved in the application of several projects of the European Union, the Federal Ministry for Economic Affairs and Energy and the Federal Ministry of Education and Research. The impressive number and the high level of her publications should also be highlighted.

I am convinced that this book offers considerable value and new insights to students, researchers, and practitioners.

Prof. Dr. Harald F. O. von Kortzfleisch

Koblenz, Germany

Abstract

Currently, many small and medium-sized enterprises (SMEs) are confronted with difficult market conditions, such as increasing globalization. In their efforts to maintain important positions in the marketplace and to compete globally, SMEs face a challenge: a lack of knowledge resources. Furthermore, the current economic crisis has weakened the financial health of many SMEs. As a consequence of these deficiencies in their internal financial resources, SMEs are unable to hire additional personnel or to invest in generating specific skills that would eliminate their knowledge-resource problems.

To address these challenges, firms are increasingly opening their boundaries and actively collaborating with outside partners to access external knowledge that will enable them to successfully innovate and remain competitive in the marketplace. To recognize the value of external knowledge, assimilate it, and apply it to commercial ends, SMEs must have absorptive capacity, that is, the distinctive capacity to absorb knowledge.

A knowledge-friendly organizational culture is the main influencing factor of absorptive capacity, in particular its acquisition capability, assimilation capability, transformation capability and exploitation capability, because people's behavior can be better coordinated through implicit values and norms than through structural coordination instruments. A knowledge-friendly organizational culture is characterized by six dimensions: trust, collaboration, openness, autonomy, learning receptivity and care.

Focusing on an organization's overall behavior, which is increasingly important in the context of external knowledge absorption, it is important to investigate in detail how a knowledge-friendly organizational culture influences SMEs' absorptive capacity. Therefore, the objective of this thesis is to develop and validate a model that allows an analysis of the relationship between organizational culture and the capabilities of absorptive capacity at the organizational level of SMEs and shows how a knowledge-oriented organizational culture should be designed to support the absorption of external knowledge.

Several theories must be consulted to develop a model of SMEs' external knowledge absorption: organizational learning, innovation, managerial cognition, knowledge-based view, dynamic capabilities and coevolution. From the perspective of these theories, hypotheses about relationships between the dimensions of organizational culture and the capabilities of absorptive capacity can be worked out and theoretically supported. These hypotheses imply a model of external knowledge absorption that is empirically manifested via quantitative methods.

The empirical analysis shows that absorptive capacity is influenced by organizational culture: acquisition capability is the only capability of absorptive capacity that is not influenced by organizational culture. Assimilation capability, transformation capability and exploitation capability are positively influenced by different dimensions of organizational culture, but by the same number of dimensions — three — of organizational culture. Therefore, organizational culture is equally important for the three capabilities of an organization.

Because this model is used both to describe the absorption of external knowledge and to explain the influence of organizational culture to derive a starting point for organizational-culture-driven control of external knowledge absorption, managers can learn methods of meeting the challenges of a knowledge-friendly organizational culture's external knowledge absorption. Additionally, this thesis explains the implications of those measures for creating a trustful, collaborative, open and learning-receptive culture.

To summarize, organizational culture generally is not positively related to the capabilities of absorptive capacity. Instead, a detailed differentiation of several dimensions of organizational culture is necessary to purposefully support external knowledge absorption in particular organizational cultures.

Contents

Foreword VII

Abstract XI

Contents XIII

List of Figures XVII

List of Tables XIX

List of Abbreviations XXII

1 The impact, Meaning and Challenges of Knowledge Absorption 1

1.1 The Problem of External Knowledge Absorption for SMEs and the Role of Organizational Culture 1

1.2 The State of the Art of Research on SMEs' External Knowledge Absorption and Organizational Culture 7
    1.2.1 The State of the Art of Research on SMEs' External Knowledge Absorption 7
    1.2.2 The State of the Art of Research on Organizational Culture 11
    1.2.3 The Research Gap in the Fields of the Absorptive Capacity and Organizational Culture of SMEs 15

1.3 Objective 18

1.4	Research Design	20
1.5	Structure	22

2 The Conceptual Principles 26

2.1	The Conceptual Principles of Absorptive Capacity	26
2.1.1	The Relationship between Absorptive Capacity and Firm Performance	28
2.1.2	Antecedents of Absorptive Capacity	30
2.1.3	Components of Absorptive Capacity	33
2.1.4	Capabilities of Absorptive Capacity	34
2.2	The Conceptual Principles of Organizational Culture	37
2.2.1	The Elements and Levels of Organizational Culture	37
2.2.2	Approaches to Organizational Culture	41
2.2.3	A Knowledge-Friendly Approach to Organizational Culture	50

3 A Model of External Knowledge Absorption 54

3.1	Absorptive Capacity in Organizational Theories	55
3.1.1	Organizational Learning	57
3.1.2	Innovation	60
3.1.3	Dynamic Capabilities	63
3.1.4	Knowledge-Based View	65
3.1.5	Managerial Cognition	67
3.1.6	Coevolution	70
3.1.7	Summary of Theories Informing Absorptive Capacity	72
3.2	The Parameters of the Model	73
3.2.1	The Construct of Absorptive Capacity	73
3.2.2	The Construct of Organizational Culture	80

3.3		The Relationship between Organizational Culture and the Capabilities of Absorptive Capacity 87
	3.3.1	The Relationship Between Organizational Culture and the Acquisition Capability 88
	3.3.2	The Relationship Between Organizational Culture and the Assimilation Capability 101
	3.3.3	The Relationship Between Organizational Culture and the Transformation Capability..111
	3.3.4	The Relationship Between Organizational Culture and the Exploitation Capability 123
3.4		The Summarized Presentation of the Model 134

4 An Empirical Analysis of the Research Models 139

4.1	Object of Study 141	
4.2	Methodology of Data Collection..142	
4.3	Operationalization of the Variables..144	
	4.3.1	Operationalization of Absorptive Capacity 144
	4.3.2	Operationalization of Organizational Culture...................148
4.4	Questionnaire 152	
4.5	Description of the Sample 153	
4.6	Descriptive Analysis 165	
4.7	Reliability Analysis..172	
4.8	Factor Analysis and Regression Analysis 176	
	4.8.1	Factor Analysis 178
	4.8.2	Regression Analysis..190

4.9   Summary of the Results 210

5   Summary, Conclusion and Outlook...219

5.1   Summary and Contribution to Management and Research 219
    5.1.1   Contributions to Management 224
    5.1.2   Contributions to Research...227

5.2   Implications for Management 230
    5.2.1   Implications for a Trustful Organizational Culture.........................232
    5.2.2   Implications for a Collaborative Organizational Culture................235
    5.2.3   Implications for an Open Organizational Culture...........................237
    5.2.4   Implications for a Learning-Receptive Organizational Culture......240

5.3   Implications for Research 242

Literature 247

List of Figures

Figure 1 The Basic Elements of a Research Process 21

Figure 2 Structure of This Thesis 25

Figure 3 A Model of Absorptive Capacity 28

Figure 4 Capabilities of Absorptive Capacity 35

Figure 5 Levels of Culture and Their Interaction 40

Figure 6 Components of a Knowledge-Friendly Organizational Culture 51

Figure 7 Dimensions of a Knowledge Culture 52

Figure 8 Theories Informing Absorptive Capacity 57

Figure 9 Detailed Aspects of the Acquisition Capability 75

Figure 10 Detailed Aspects of the Assimilation Capability 76

Figure 11 Detailed Aspects of the Transformation Capability 78

Figure 12 Detailed Aspects of the Exploitation Capability 80

Figure 13 Importance of Trust 81

Figure 14 Research Model of the Acquisition Capability 99

Figure 15 Research Model of the Assimilation Capability 109

Figure 16 Research Model of the Transformation Capability 121

Figure 17 Research Model of the Exploitation Capability 131

Figure 18 Summary of Hypotheses H_1 through H_{24} 136

Figure 19 Structure of the Empirical Analysis 140

Figure 20 Construct of Absorptive Capacity 148

Figure 21 Construct of Organizational Culture 151

Figure 22 Structure of the Questionnaire 153

Figure 23 Histogram: 'In Our Company, Problems Rarely Arise, Because We Have The Skills That Are Required For Our Jobs' (Autonomy) 171

Figure 24 Histogram: 'In Our Daily Working Environment, Working Procedures Are Reviewed And Improved' (Learning Receptivity) 172

Figure 25 An Absorptive-Capacity-Supporting Organizational Culture 231

List of Tables

Table 1 External Knowledge Sources 31

Table 2 Approaches to Organizational Culture (I) 44

Table 3 Approaches to Organizational Culture (II) 45

Table 4 Approaches to Organizational Culture (III) 46

Table 5 Approaches to Organizational Culture (IV) 47

Table 6 Approaches to Organizational Culture (V) 48

Table 7 Closed Innovation Principles Versus Open Innovation Principles 61

Table 8 Theories Informing the Acquisition Capability 101

Table 9 Theories Informing the Assimilation Capability 110

Table 10 Theories Informing the Transformation Capability 122

Table 11 Theories Informing the Exploitation Capability 132

Table 12 Theories Informing the Capabilities of Absorptive Capacity 135

Table 13 Summary of the Operationalization of Absorptive Capacity 147

Table 14 Summary of the Operationalization of Organizational Culture (I) 150

Table 15 Summary of the Operationalization of Organizational Culture (II) 151

Table 16 Field Report 154

Table 17 Distribution by Industrial Sector 155

Table 18 Firm Size 156

Table 19 Firm Sales 157

Table 20 Respondents' Business Units 158

Table 21 Respondents' Job Positions 158

Table 22 Job Position * Business Unit Cross Tabulation (I) 160

Table 23 Job Position * Business Unit Cross Tabulation (II) 161

Table 24 Descriptive Statistics of the Absorptive Capacity (I) 166

Table 25 Descriptive Statistics of the Absorptive Capacity (II) 167

Table 26 Descriptive Statistics of Organizational Culture (I) 168

Table 27 Descriptive Statistics of Organizational Culture (II) 168

Table 28 Selected Recommended Reliability Levels 173

Table 29 Reliability Statistics of the Absorptive Capacity 175

Table 30 Reliability Statistics of the Organizational Culture 176

Table 31 Evaluation of Levels of KMO 180

Table 32 Factor Analysis of the Absorptive Capacity 182

Table 33 Interpretation of Correlation Coefficients 183

Table 34 Correlations of the Absorptive Capacity 185

Table 35 Factor Analysis of the Organizational Culture (I) 185

Table 36 Factor Analysis of the Organizational Culture (II) 187

Table 37 Correlations of the Organizational Culture 188

Table 38 Coefficients of the Acquisition Capability 192

Table 39 Overview of Hypotheses H_1 through H_6 194

Table 40 Coefficients of the Assimilation Capability 197

Table 41 Overview of the Hypotheses H_7 through H_{12} 199

Table 42 Coefficients of the Transformation Capability 201

Table 43 Overview of the Hypotheses H_{13} through H_{18} 205

Table 44 Coefficients of the Exploitation Capability 206

Table 45 Overview of the Hypotheses H_{19} to H_{24} 210

Table 46 Summary of the Examination of Hypotheses H_1 through Hypotheses H_{24} 212

Table 47 Positive Relationships for Each Capability of Absorptive Capacity 214

List of Abbreviations

ACAP	absorptive capacity
B	regression coefficient
BU	business unit
cf.	confer
e.g.	exempli gratia
f.	following
ff.	and the following ones
ICT	information and communication technology
IJV	international joint venture
KMO	Kaiser-Meyer-Olkin
K-S	Kolmogorov-Smirnov
min	minutes
ML	maximum likelihood factor analysis
MNC	multinational corporation
MSA	Measure of Sampling Adequacy
n.s.	not significant
NIH	Not Invented Here
OECD	Organization for Economic Co-operation and Development
p.	page
PAF	principal axis factoring
PCA	principal component analysis
R&D	research and development
R^2	coefficient of determination
s	seconds
SE B	standard error of the regression coefficient
SME	small and medium-sized enterprises
std.	Standard
U.K.	United Kingdom
U.S.	United States

List of Abbreviations

URL Uniform Resource Locator
β standardized regression coefficient

1 The Impact, Meaning and Challenges of Knowledge Absorption

1.1 The Problem of External Knowledge Absorption for SMEs and the Role of Organizational Culture

At the present time, many small and medium-sized enterprises (SMEs) are confronted with difficult market conditions. These conditions involve, for example, increasing globalization, greater competitive intensity, advances in information and communication technology (ICT) and changes in organizational structures (cf. Argote, McEvily & Reagans 2003a, p. v, cf. Ebrahim, Ahmed & Taha 2010, p. 917, cf. Hall 2003, p. iii, cf. Picot, Reichwald & Wigand 2003, p. 2 ff.). As a consequence, SMEs must be able to quickly and flexibly respond to market changes to stay competitive.

In their efforts to maintain an important position in the marketplace and to conquer global competition, SMEs face a major resource problem:

First, SMEs are challenged by a lack of knowledge resources. This issue is especially important because knowledge is a prerequisite for innovation and firms can generate long-term competitive advantages through innovation. To innovate, SMEs have traditionally pursued relatively closed innovation processes using their own resources and have focused on internal research and development (R&D), with most innovation-related activities conducted inside organizational boundaries (cf. Lichtenthaler & Lichtenthaler 2012, p. 154). As a result, those firms'[1] critical knowledge was primarily developed and applied in-

---

[1] The terms 'firm', 'business', 'company', 'enterprise' and 'organization' are used synonymously throughout this thesis.

house. The extent of interorganizational knowledge transfer[2] was relatively limited.

Second, the current economic crisis has weakened many SMEs' financial health. An investigation by the *OECD*[3] *Centre for Entrepreneurship, SMEs and Local Development* (2009) has provided many reasons to explain why SMEs are generally more vulnerable in times of crisis: "it is more difficult for them to downsize as they are already small; they are individually less diversified in their economic activities; they have a weaker financial structure (i.e. lower capitalisation); they have a lower or no credit rating; they are heavily dependent on credit and they have fewer financing options" (OECD Centre for Entrepreneurship, SMEs and Local Development 2009, p. 6; own citation[4]). That investigation has indicated that deficiencies in internal financial resources are a major liability of SMEs.

Third, in terms of management of new technologies, SMEs have fewer resources and expertise than do large enterprises (cf. Lawrence 2010, p. 38 ff.). This additional lack of human resources (cf. Welsh & White 1981, p. 18) and specialized knowledge confirms that SMEs are more vulnerable than large enterprises.

In summary, SMEs lack those resources that are required to fulfill processes that generate long-term competitive advantages (cf. Vanhaverbeke 2012, p. 9). Therefore, resources are SMEs' main challenge. To address this resource challenge, important practical changes in the business landscape are now occurring because firms have become increasingly focused on interorganizational

---

[2] "It is striking to note that several authors refer to the act of knowledge transfer as a form of translation. They appear to mean this as a metaphor, but actually translation in this respect is better conceived of as an analogy" (Holden & von Kortzfleisch 2004, p. 129).

[3] 'OECD' is an abbreviation of 'Organisation for Economic Co-operation and Development'.

[4] 'Own citation' means that, e.g., with regard to formatting, bullets were changed or replaced by semicolons and italic font was changed to normal font.

cooperation that enables flexible interdependencies and co-development among various firms (cf. Ding, Akoorie & Pavlovich 2009, p. 48). Through these interorganizational relationships, firms acquire knowledge, skills and resources that complement their internal R&D but that they normally cannot develop independently (cf. Albers 2010, p. 206). Alliances, acquisitions and joint ventures are examples of such interorganizational cooperation,[5] and the number has increased in recent decades (cf. Ding et al. 2009, p. 47, cf. Powell, Koput & Smith-Doerr 1996, p. 116). SMEs open up their innovation processes and actively collaborate with external partners throughout the innovation process to access knowledge that exists beyond their own boundaries to successfully innovate and remain competitive in the marketplace.

To recognize the value of knowledge from external sources, assimilate it, and apply it to commercial ends, SMEs need a distinctive capacity to absorb knowledge. This ability to explore external knowledge, called 'absorptive capacity' (cf. Lichtenthaler & Lichtenthaler 2009, p. 1319), can be a source of a firm's competitive advantage because the differences in firms' utilization of organizational resources and capabilities impact their performance (cf. Zahra & George 2002, p. 195)[6].

To cope with the challenge of the lack of resources, the topic of knowledge and its management have evolved in recent years into essential issues in both business practice and theory (e.g., Argote et al. 2003a, p. v, cf. Becker & Knudsen 2006, p. 4, Moffett, McAdam & Parkinson 2002, p. 237, Scholl, König, Meyer & Heising 2004, p. 19). They are discussed in various disciplines such as organizational behavior, information systems, strategic management, psychology and sociology (cf. Argote, McEvily & Reagans 2003b, p. 571, Tang, Xi & Ma 2006, p. 796). In economics, as a fundamental science, the

---

[5] Further kinds of interorganizational cooperation and sources of external knowledge are explained in section 2.1.2.
[6] *Zahra & George* (2002, p. 195), referencing *Barney* (1991), *Spender* (1996) and *Teece, Pisano & Shuen* (1997).

traditional three factors of production, namely, land, labor and capital goods, have been supplemented by a fourth factor of production: 'knowledge' (cf. Evers 2002, p. 2 ff., cf. Tang 2005, p. 1 ff., cf. Westeren 2006, p. 1).

The burgeoning interest in knowledge is reflected by the increasing volume of research. From a historical perspective, *Penrose* (1959) emphasizes the importance of knowledge for the activities of an enterprise during the course of his investigations of corporate growth. He concluded that "the significance of resources to a firm and the productivity they can yield [are] functions of knowledge" (Penrose 1959, p. 77). This view influences the relevance of absorptive capacity because absorptive capacity is "the ability of a firm to recognize the value of new, external information, assimilate it, and apply it to commercial ends" (Cohen & Levinthal 1990, p. 128). Therefore, absorptive capacity is essential to develop and increase the knowledge base of a firm (cf. Volberda, Foss & Lyles 2010, p. 935). *Kogut & Zander* (1992) consider a firm's knowledge base as the main determinant of competitive advantage. In their view, the ability of a firm to create and transfer knowledge efficiently within an organizational context is the central competitive dimension (cf. Kogut & Zander 1992, p. 384). One year later, operating from the perspective that enterprises exist to internalize markets, *Kogut & Zander* (1993) developed the view that "firms [...] serve as efficient mechanisms for the creation and transformation of knowledge into economically rewarded products and services" (Kogut & Zander 1993, p. 627) and once again supported the relevance of absorptive capacity.

Argote, Ingram, Levine & Moreland (2000) propose that businesses with successful knowledge management and the ability to transfer knowledge effectively between their units are more productive and more likely to survive (cf. Argote et al. 2000, p. 1), and therefore, they support the role of knowledge and knowledge management for generating competitive advantages. *Liebowitz* (1999) specifies six influencing factors that lead to successful knowledge management and thereby ensure long-term competitive advantages: (1) knowledge management strategy; (2) knowledge management infrastructure; (3) knowledge ontologies and repositories; (4) knowledge management systems and tools; (5)

incentives to motivate employees to share knowledge; and (6) a supportive culture for knowledge management (cf. Liebowitz 1999, p. 37 f.). *Davenport, De Long & Beers* (1998) also investigate influencing factors for successful knowledge management and specify eight influencing factors: (1) a link to economic performance or industry value; (2) technical and organizational infrastructure; (3) a standard, flexible knowledge structure; (4) a knowledge-friendly culture; (5) a clear purpose and language; (6) changes in motivational practices; (7) multiple channels for knowledge transfer; and (8) senior management support (cf. Davenport et al. 1998, p. 50).

Both *Liebowitz* (1999) and *Davenport et al.* (1998), among others, emphasize a supportive and knowledge-friendly organizational culture as an influencing factor for knowledge management. *Liebowitz* (1999) states that a 'supportive culture for knowledge management' is an influencing factor (cf. Liebowitz 1999, p. 38). *Davenport et al.* (1998) state that 'knowledge-friendly culture' is an influencing factor (cf. Davenport et al. 1998, p. 50). In the extant literature, various other authors confirm this consideration and even describe organizational culture as a key success factor for the management and transfer of knowledge (e.g., cf. Ford & Chan 2002, p. 8, cf. Glisby & Holden 2003, p. 29 ff., cf. King 2007, p. 226 f., cf. Moffett et al. 2002, p. 238).

As a key success factor for the management and transfer of knowledge[7], a knowledge-friendly organizational culture is very important for competitiveness, especially to SMEs. Although SMEs face several constraints with respect to differentiating their products and services and changing aspects of their business model (cf. Vanhaverbeke 2012, p. 9), the establishment of a

---

[7] Of course, the other influencing factors for knowledge transfer are important, too. Nevertheless, it has been deliberately decided, in the context of this thesis, to investigate the most difficult tangible and (as presented by many authors) particularly important factor: organizational culture. In anticipation of the statement of reasons of the research gap in section 1.2.3, it is already possible to note that with respect to the organizational culture, most sustainable research needs exist.

knowledge-friendly organizational culture is one possible method by which SMEs can distinguish themselves in the market and take advantage of their specific competitive advantages (cf. Spieth 2009, p. 2). In addition to providing this advantage of SME-specific competitiveness, a strong organizational culture is highly important to the long-term survival of an SME because processes in dynamic, uncertain environments can be better coordinated through implicit values and norms better than through structural coordination instruments (cf. Spieth 2009, p. 2). The reason why these processes can be coordinated through implicit values and norms is that "culture is believed to influence the knowledge-related behaviors of individuals, teams, organizational units and overall organizations because it importantly influences the determination of which knowledge it is appropriate to share, with whom and when" (King 2007, p. 226; own citation). This potential of organizational culture should be exploited by SMEs in particular because these companies usually do not have the resources to comprehensively establish explicit structural coordination instruments or motivation rules. In addition, SMEs give up an advantage, which can be seen in less-bureaucratized structures: strategic flexibility.

In general, despite the fact that SMEs are considered "the engines of global economic growth" (Ebrahim et al. 2010, p. 916) and thus to be indispensable for the economy and employment of any country, the current research on SMEs has remained rather limited (van de Vrande, de Jong, Vanhaverbeke & de Rochemont 2008 p. 3, De Jong & Marsili 2005, p. 2, Masse & Testa 2008, p. 394). *Flatten, Greve & Brettel* (2011b) highlight that the effects of absorptive capacity with regard to SMEs have been only marginally addressed by the literature (cf. Flatten et al. 2011b, p. 139) and in particular, the relationship between organizational culture and absorptive capacity, especially in the context of SMEs, has not been adequately addressed by existing investigations, as will be elaborated in section 1.2.

To clarify the need for research on how organizational culture influences the absorptive capacity of SMEs, section 1.2 discusses state of the art research on SMEs' external knowledge absorption and organizational culture.

1 The Impact, Meaning and Challenges of Knowledge Absorption

1.2 The State of the Art of Research on SMEs' External Knowledge Absorption and Organizational Culture

Here, the state of the art of research on the topic of SMEs' external knowledge absorption is discussed in section 1.2.1 and state of the art research on the topic of organizational culture is discussed in section 1.2.2. Subsequently, the research gap with respect to both of these research fields is identified and in section 1.2.3, the central research question of this thesis is formulated in greater detail.

1.2.1 The State of the Art of Research on SMEs' External Knowledge Absorption

In recent years, the volume of research on knowledge transfer has become more important, and this topic has been increasingly discussed in the literature (e.g., cf. Argote et al. 2003a, p. v, Becker & Knudsen 2006, p. 5, Moffett et al. 2002, p. 237, Scholl et al. 2004, p. 19). Knowledge transfer is an essential aspect of knowledge management. *Minbaeva, Pedersen, Björkman, Fey & Park* (2003) define knowledge transfer as "a process that covers several stages starting from identifying the knowledge over the actual process of transferring the knowledge to its final utilization by the receiving unit" (Minbaeva et al. 2003, p. 587). Enterprises use a variety of different mechanisms for coordinating knowledge transfer stages when they open up their innovation processes and actively acquire and explore external knowledge from sources such as interorganizational relationships, including R&D consortia, strategic alliances, joint ventures, patents, interactions with suppliers and customers and staff transfers, to successfully innovate and remain competitive in the market[8].

---

[8] External knowledge resources are described in detail in section 2.1.2.

To absorb acquired knowledge from external sources, firms need to have adequate absorptive capacity. Based on *Cohen & Levinthal*'s (1990) original definition, absorptive capacity is "the ability of a firm to recognize the value of new, external information, assimilate it, and apply it to commercial ends" (Cohen & Levinthal 1990, p. 128)[9]. This ability to absorb external knowledge is largely a function of a firm's level of prior related knowledge (cf. Cohen & Levinthal 1990, p. 128) and relates to the concept of inward technology transfer[10]. Inward technology transfer, a one-way transfer of knowledge from the environment to a firm, allows firms to acquire all of the external knowledge that is available on the market (cf. Lichtenthaler & Lichtenthaler 2012, p. 156). In contrast to inward technology transfer, firms can commercialize everything in their portfolio via outward technology transfer (cf. Lichtenthaler & Lichtenthaler 2012, p. 156). For a successful outward technology transfer, a sufficient ability to externally exploit knowledge, called desorptive capacity, is needed (cf. Lichtenthaler & Lichtenthaler 2009, p. 1322). Desorptive capacity refers to a "firm's ability to externally exploit knowledge" (Lichtenthaler & Lichtenthaler 2009, p. 1322). It is a determinant of the potential volume of technology transfer based on a firm's technology portfolio (cf. Lichtenthaler & Lichtenthaler 2012, p. 157)[11].

---

[9] *Zahra & George* (2002) reconceptualize Cohen & Levinthal's (1990) construct of absorptive capacity and define absorptive capacity as "a set of organizational routines and processes by which firms acquire, assimilate, transform, end exploit knowledge to produce a dynamic organizational capability" (Zahra & George 2002, p. 186). The conceptual principles of absorptive capacity are explained in more detail in section 2.1.

[10] The terms 'knowledge transfer' and 'technology transfer' are used synonymously throughout this thesis.

[11] *Lichtenthaler & Lichtenthaler* (2012) note that research and practice have primarily focused on the recipient's absorptive capacity and not on the sender's desorptive capacity (cf. Lichtenthaler & Lichtenthaler 2012, p. 165). They give at least two reasons for this emphasis: "First, the technology recipient is normally not only involved in the technology transfer activities. Instead, it has to subsequently integrate and finally apply the technological knowledge, e.g., in new products. Accordingly, the recipient's role may be more prominent although not necessarily more important in in-

1 The Impact, Meaning and Challenges of Knowledge Absorption

Without the ability to explore external knowledge (cf. Lichtenthaler & Lichtenthaler 2009, p. 1319), SMEs cannot benefit from better ideas or ways of performing operations — they cannot use external knowledge sources. Because of this need for absorptive capacity, it has received substantial attention in both research and practice (e.g., cf. Flatten, Engelen, Zahra & Brettel 2011a, p. 98, cf. Lane, Koka & Pathak 2006, p. 833, cf. Lichtenthaler & Lichtenthaler 2012, p. 155, cf. Volberda, Foss & Lyles 2009, p. 1, cf. Zahra & George 2002, p. 185).

Cohen & Levinthal (1990) argue that absorptive capacity is critical to an enterprise's innovative capabilities and that it is dependent on an enterprise's level of prior related knowledge.

Szulanski (1996) adds that a recipient's lack of absorptive capacity is a major barrier to internal knowledge transfer. *Tsai* (2001) investigates the influence of the interaction between the absorptive capacity and the network position on the innovation and performance of business units. His research demonstrates that a high absorptive capacity is associated with a better chance for the successful application of new knowledge toward commercial ends which in turn, produce more innovations and better business performance (cf. Tsai 2001, p. 1003).

Liao, Welsch & Stoica (2003) state that SMEs with a high level of absorptive capacity are more proactive for two reasons (cf. Liao et al. 2003, p. 69): First, the "development of absorptive capacity in certain areas permits an SME to better appreciate, understand, and evaluate the merit of environmental signals" (Liao et al. 2003, p. 69). Second, absorptive capacity "enables an SME to

---

terorganizational technology transactions. Second, the performance of one single technology transaction can often be better evaluated from the perspective of the technology recipient who subsequently exploits the technological knowledge" (Lichtenthaler & Lichtenthaler 2012, p. 165). As mentioned in section 1.1, resources are the main challenge for SMEs. SMEs open up their innovation processes and actively collaborate with external partners throughout the innovation process to integrate and finally, apply technological knowledge from external sources to new products and services. Therefore from an SME's point of view, absorptive capacity is of dominant importance for their ability to innovate to remain competitive in the market.

more readily accumulate what additional knowledge it needs to exploit any critical knowledge that may become available" (Liao et al. 2003, p. 69).

Zhao & Anand (2009) investigate the influence of absorptive capacity on knowledge transfer at the individual and collective levels and empirically validated the distinction between individual and collective teaching activities and absorptive capacity. They discovered that collective-level mechanisms, e.g., collective teaching, are more effective for the transfer of collective and individual knowledge in comparison to individual teaching (cf. Zhao & Anand 2009, p. 959).

Lichtenthaler (2009) investigates interfirm discrepancies in profiting from external knowledge and identifies technological and market knowledge as two critical components of prior knowledge in organizational learning processes (cf. Lichtenthaler 2009, p. 822).

Lichtenthaler & Lichtenthaler (2009) investigate interfirm heterogeneity in knowledge and alliance strategies, organizational boundaries and innovation performance; through this examination, these researchers identify a firm's knowledge capacities that are critical for managing internal and external knowledge (cf. Lichtenthaler & Lichtenthaler 2009, p. 1315).

Lichtenthaler & Lichtenthaler (2012) discuss the implications of desorptive capacity for outward technology transfer performance and consider the potential interactions of absorptive capacity and desorptive capacity in interorganizational technology transfer. The presented interdependencies between absorptive and desorptive capacity deepen the understanding of the success or failure of interorganizational technology transfer.

In summary, researchers have shown that absorptive capacity is related, inter alia, to innovation and learning (cf. Cohen & Levinthal 1990), knowledge transfer in intraorganizational networks (cf. Tsai 2001), growth (cf. Liao et al. 2003), collective and individual knowledge (cf. Zhao & Anand 2009), technological and market knowledge (cf. Lichtenthaler 2009), knowledge capacities (cf. Lichtenthaler & Lichtenthaler 2009) and desorptive capacity (cf. Lichtenthaler & Lichtenthaler 2012) and have referred to the construct of absorptive capacity in more than 10,000 published papers, chapters and books (cf. Lewin, Massini & Peeters 2011, p. 81). Nevertheless, the relationship between organizational culture and the absorption of external knowledge has not been comprehensively addressed in the extant literature. The broad topic of organizational culture as an influencing factor of the absorption of external knowledge, in particular the acquisition capability, assimilation capability, transformation capability and exploitation capability, has been largely neglected — especially in the context of SMEs.

The research gap with respect to organizational culture as an influencing factor of the capacity of SMEs to absorb external knowledge and the central research question of this thesis are formulated in greater detail in section 1.2.3, after the discussion of state of the art research on organizational culture, which is found in the following section.

1.2.2 The State of the Art of Research on Organizational Culture

The notion of organizational culture has been important to both practice and research since the beginning of the 1980s, and interest in the phenomenon of organizational culture has greatly increased in recent decades (cf. Barley 1983, p. 393, cf. O'Reilly 1991, p. 487, cf. Smircich 1983, p. 339). Many investigations from the 1980s and the early 1990s are focused on the management of organizational culture and the influence of organizational culture on the success of an enterprise (e.g., Deal & Kennedy 1982, Denison 1990, Kotter & Heskett 1992, Peters & Watermann 1982). *Koberg & Chusmir* (1987) state that "prior research suggests that favorable work outcomes are a function of how well people's needs or personalities are matched by a number of work-environment variables" (Koberg & Chusmir 1987, p. 397) and note that organizational culture is one of these environmental variables.

Organizational culture conveys a sense of identity to organization members, encourages a commitment that extends beyond the self, enhances the stability of the social system, provides premises for decision-making that guide and shape behaviors (cf. Smircich 1983, p. 345 f.) and is an "essential factor, which not only guarantees a successful knowledge management but also influences a successful knowledge transfer" (Girdauskienė & Savanevičienė 2007, p. 36; own formatting). In the extant literature, various other authors confirm this consideration and describe organizational culture as a key influencing factor for the management and transfer of knowledge (e.g., cf. Ford & Chan 2002, p. 8, cf. Glisby & Holden 2003, p. 29 ff., cf. King 2007, p. 226 f., cf. Moffett et al. 2002, p. 238).

Lemken, Kahler & Rittenbruch (2000) analyze the influences of the technical and organizational aspects of knowledge processes in virtual organizations in the context of organizational culture. They discover that cooperation of all members and at all levels of an organization is necessary for sustained knowledge management (cf. Lemken et al. 2000, p. 9). Such cooperation becomes possible when all members have common goals, values and procedures and work in an trustful environment because knowledge sharing requires trust (cf. Lemken et al. 2000, p. 9).

Holden (2001) addresses the cross-cultural dimensions of knowledge management; in particular, he provides a general review of knowledge management ideas and examines potentially important cross-cultural implications and knowledge management failures. His findings suggest that it is the key task of knowledge management to foster and continually sophisticate collaborative, cross-cultural learning and therefore, it is very important that what people learn from each other is not the essence of the cross-cultural challenge, it is how they learn (cf. Holden 2001, p. 155).

Bhagat, Kedia, Harveston & Triandis (2002) investigate the ways in which organizational culture influences knowledge transfers among organizations that are located in dissimilar cultural contexts. They state that organizational culture has a significant influence on the effectiveness of cross-border knowledge transfers (cf. Bhagat et al. 2002, p. 206 & 218). Furthermore, they claim that the effectiveness of cross-border knowledge transfers is influenced by the types of knowledge that are being transferred, the nature of transacting cultural patterns and cognitive styles (cf. Bhagat et al. 2002, p. 206).

Ford & Chan (2002) examine relationships between cross-cultural differences and knowledge transfer within an international subsidiary and reveal that language differences can create knowledge blocks and that cross-cultural differences can explain the direction of knowledge flows within an international subsidiary (cf. Ford & Chan 2002, p. 3).

Based on the finding that a number of critical factors for knowledge management have been highlighted in the literature, e.g., the macroenvironment along with technological, informational, cultural and people factors, *Moffett et al.* (2002) claim that the relationship between the cultural and technological aspects of knowledge management is limited by empirical research (cf. Moffett et al. 2002, p. 237). Therefore, they develop a conceptual model for knowledge management implementation that can be used for several types of organizations to show how organizations can systematically integrate the cultural and technological aspects of knowledge management.

Lucas (2006) investigate the influence of culture on knowledge transfer within multinational corporations (MNCs) with subsidiaries that operate in heterogeneous national cultures. He suggests that managers should pursue knowledge transfer activities cautiously and points out that resistance to change and sharing must be carefully managed (cf. Lucas 2006, p. 257). Furthermore, he determines that the success of knowledge transfer efforts depends on the parties' cultural alignment and that when this cultural alignment is missing, success is highly dependent upon directives and support from a home office (cf. Lucas 2006, p. 257).

Weissenberger-Eibl & Spieth (2006) assess the effects of organizational culture on knowledge transfer, reveal cultural constraints and discuss the implications of improving knowledge transfer across cultural boundaries. They point out two objectives of considering the impact of culture on knowledge transfer in an appropriate manner are important for successful companies: "1) sensitization for corporate culture and training for acculturation and 2) using instruments and applying methods to enhance the effectiveness of the transfer of implicit knowledge" (Weissenberger-Eibl & Spieth 2006, p. 72)[12]. They highlight the importance of these objects, stating that "managing implicit knowledge as well as cross-cultural work will be core competencies of successful companies in [the] future" (Weissenberger-Eibl & Spieth 2006, p. 72; own citation).

King (2007) investigate various concepts and levels of organizational culture that are useful to research and practice in knowledge management and show that organizational culture is both an antecedent and an outcome of knowledge management (cf. King 2007, p. 227).

Girdauskienė & Savanevičienė (2007) identify different features of culture, e.g., empowerment, motivation and mistake tolerance, and discussed the ways in which these features influence the effectiveness of knowledge transfer (cf. Girdauskienė & Savanevičienė 2007, p. 40).

Spieth (2009) investigates the influence of organizational culture on knowledge transfer at the individual level in enterprises and examines the ways in which a knowledge-oriented culture can be designed to foster successful knowledge transfers.

In summary, according to a plethora of sources organizational culture is a critical factor for knowledge management; in fact, organizational culture may be the most critical determinant. Although the role of organizational culture for knowledge management has been recognized in research and in practice, how-

---

[12] The style of English was changed from British to American.

ever, the existing literature provides only initial answers regarding the relationship between organizational culture and knowledge management — especially with regard to absorptive capacity. The previously mentioned studies by *Lemken et al.* (2000), *Holden* (2001), *Bhagat et al.* (2002), *Ford & Chan* (2002), *Moffett et al.* (2002), *Lucas* (2006), *Weissenberger-Eibl & Spieth* (2006), *King* (2007), *Girdauskienė & Savanevičienė* (2007) and *Spieth* (2009) either treat this topic in a cursory manner or only attempt to examine those influences of organizational culture that are produced by selected aspects of several dimensions of organizational culture on knowledge management and not on absorptive capacity in detail. This has resulted in a research gap in the fields of the absorptive capacity and organizational culture of SMEs that is discussed in the following section 1.2.3.

1.2.3 The Research Gap in the Fields of the Absorptive Capacity and Organizational Culture of SMEs

The current research on SMEs has remained rather limited (e.g., van de Vrande et al. 2008 p. 3, De Jong & Marsili 2005, p. 2, Massa & Testa 2008, p. 394), especially with regard to absorptive capacity, although the number of investigations that examine various aspects of absorptive capacity generally has significantly increased in recent decades (e.g., Flatten et al. 2011a, p. 98, Lane et al. 2006, p. 833, Volberda et al. 2009, p. 1). Absorptive capacity has primarily been studied in the context of large, multinational enterprises, which typically contain sizeable internal R&D departments. The management of SMEs' inward knowledge transfer is a rather specific process, and lessons from large enterprises' external knowledge absorption are not readily transferable to the SME context. The reason is that the size of SMEs is accompanied by resource poverty — as highlighted in section 1.1 — that distinguishes them from large firms and makes different types of management approaches necessary (cf. Welsh & White 1981, p. 2). Due to their size and resource poverty, SMEs can be expected to have a lower level of absorptive capacity, a less-developed approach toward absorptive capacity, a lack of awareness of the relevance of absorptive capacity and no sound understanding of the processes of the acquisition, assimilation, transformation and exploitation of knowledge (cf. Ndiege et al. 2012, p. 6).

Organizational culture as another prominent research topic has received substantial attention in both practice and research since the beginning of the 1980s (e.g., Barley 1983, p. 393, cf. O'Reilly 1991, p. 487, cf. Smircich 1983, p. 339). Unfortunately, the results of the investigations described in section 1.2.2 are difficult to compare to each other because they involve diverse definitions of terms related to organizational culture[13]. For example in the extant literature, several terms are used synonymously for organizational culture, such as corporate culture and enterprise culture, and altogether more than 160 definitions of culture exist in the fields of anthropology, sociology and psychology (cf. Girdauskienė & Savanevičienė 2007, p. 39). Furthermore, the broad topic of organizational culture as an influencing factor of the absorption of external knowledge has been largely neglected — especially with regard to the several capabilities of absorptive capacity and especially in the context of SMEs and in detail.

In summary, although the construct of absorptive capacity has been given considerable academic attention over the last years (e.g., Flatten et al. 2011a, p. 98, Lane et al. 2006, p. 833, Volberda et al. 2009, p. 1), there is still a research area which is underdeveloped, namely, absorptive capacity in the SME context (e.g., cf. Liao et al. 2003, p. 64). *Wong & Aspinwall* (2005) highlight the scarcity of empirical studies that have investigated factors critical for the absorption of knowledge in the particular business sector of SMEs (cf. Wong & Aspinwall 2005, p. 65). Moreover, studies on knowledge management and the absorption of knowledge have only attempted to examine the influences of organizational culture that are produced by selected aspects of several dimensions of organizational culture, not absorptive capacity in detail (e.g., Lemken et al.

---

[13] Therefore, there have also been different approaches to the theoretical constructs of organizational culture, which are discussed in section 2.2.2.

2000, Holden 2001, Bhagat et al. 2002, Ford & Chan 2002, Moffett et al. 2002, Lucas 2006, Weissenberger-Eibl & Spieth 2006, King 2007, Girdauskienė & Savanevičienė 2007, Spieth 2009). Organizational culture and absorptive capacity have not been investigated, nor has there been a differentiation between the several dimensions of organizational culture and the capabilities of absorptive capacity. The literature has not investigated how the several dimensions of organizational culture are related to the acquisition capability, assimilation capability, transformation capability and exploitation capability or how a knowledge-oriented organizational culture should be designed to support the absorption of external knowledge. To close this gap, the following **central research question** is investigated in this thesis:

How is organizational culture related to the acquisition capability, assimilation capability, transformation capability and exploitation capability of a SME and how should a knowledge-oriented organizational culture be designed to support the absorption of external knowledge?

This central research question implies a research problem that can be subdivided into two problem areas:

1. The first problem area includes unresolved questions with respect to the parameters of a model of SMEs' external knowledge absorption. It is unclear which dimensions of the prevailing organizational culture influence the several capabilities of SMEs' absorptive capacity; moreover, it is also uncertain which indicators can be used to measure the dimensions of a knowledge-friendly organizational culture.
2. The second problem area that is addressed by this thesis involves unresolved questions with respect to the relationship between the dimensions of organizational culture and the capabilities of absorptive capacity. This understanding is important for designing a model that permits an assessment of the role of organizational culture in ensuring the absorption of external knowledge.

This thesis addresses the above-identified problem areas and seeks to answer the central research question. Concretized research problems serve as the starting point for the formulation of the hypotheses of this thesis in section 3.3.

Based on the presentation of the problem in section 1.1 and of the research gap in this section 1.2.3, the objective of this thesis is identified in the following section 1.3.

1.3 Objective

The presented problems regarding SMEs' external knowledge absorption represent issues that have been insufficiently studied. *The **overall objective** of this thesis is to develop and validate a model that*

- *allows an analysis of the relationship between organizational culture and the capabilities of absorptive capacity at the organizational level of SMEs and*
- *shows how a knowledge-oriented organizational culture should be designed to support the absorption of external knowledge.*

To achieve this objective of this thesis, a model of the external knowledge absorption of organizational-driven SMEs must be developed. To develop this model of SMEs' external knowledge absorption, several theories must be consulted[14]. *Cohen & Levinthal* (1989 & 1990) position absorptive capacity as a key construct in the literature on organizational learning and innovation and, in subsequent publications, lay the groundwork for theoretical contributions, constructs and implications (cf. Volberda et al. 2010, p. 932 ff.). However, the absorptive capacity theme overlaps with themes other than organizational learning and innovation, such as managerial cognition, the knowledge-based view, dy-

---

[14] The organizational theories that inform absorptive capacity are explained in detail in section 3.1.

1 The Impact, Meaning and Challenges of Knowledge Absorption 19

namic capabilities and coevolution. Therefore, another topic of discussion is how various theories in the organization field are related to absorptive capacity, for the purpose of identifying theories suitable to explain success in external knowledge absorption and the relationship between organizational culture and absorptive capacity. Next, organizational culture must be analyzed with respect to the construct of absorptive capacity and therefore, the critical dimensions of organizational culture for absorptive capacity must be identified.

From the theoretical perspective, several hypotheses about the relationships among the dimensions of organizational culture and the capabilities of absorptive capacity are worked out and theoretically supported. The hypotheses imply a model of external knowledge absorption that must be empirically manifested via an empirical analysis. Next, the design parameters of a concept for external knowledge absorption, in particular in the context of organizational culture, including recommendations for action by SMEs to maximize the absorption of external knowledge, are identified and explained, and a conclusion that includes, inter alia, further research is provided.

The **contribution** of this thesis is the identification, theory-based analysis and empirical investigation of the relationship between the dimensions of organizational culture and the capabilities of the absorptive capacity of SMEs. This is accomplished through both a theory-based analysis and an empirical investigation. The findings are transferred to implications for management to purposefully support external knowledge absorption with a knowledge-friendly organizational culture. They include recommendations for SME action to maximize the prospects of success in external knowledge absorption.

Section 1.4 describes the effort that must be put into the intended research design.

1.4 Research Design

As a starting point in developing a thesis, considerable effort must be put into answering several questions that relate, in particular, to research design:

- "What *methods* do we propose to use?
- What *methodology* governs our choice and use of methods?
- What *theoretical perspective* lies behind the methodology in question?
- What *epistemology* informs this theoretical perspective?" (Crotty 2003, p. 2)

These four questions address methods, methodology, theoretical perspective and epistemology. They are the basic elements of any research process (cf. Figure 1; cf. Crotty 2003, p. 2).

Following the approach to the four basic elements of the research process set forth by *Crotty* (2003), methods are "the techniques or procedures used to gather and analyze data related to some research question or hypothesis" (Crotty 2003, p. 3)[15] and methodology is "the strategy, plan of action, process or design lying behind the choice and use of particular methods and linking the choice and use of methods to the desired outcomes" (Crotty 2003, p. 3). This thesis used a questionnaire and survey research as the methodology that governed the choice and use of the questionnaire as method. The method and methodology are described in section 4.

---

[15] The style of English was changed from British to American.

1 The Impact, Meaning and Challenges of Knowledge Absorption   21

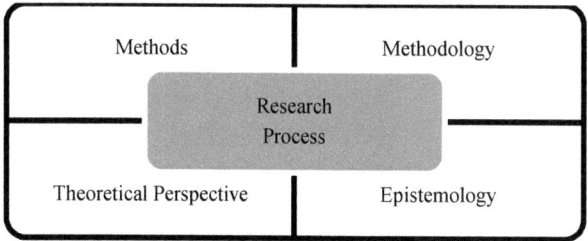

Figure 1 The Basic Elements of a Research Process[16]

The theoretical perspective is "the philosophical stance informing the methodology and thus providing a context for the process and grounding its logic and criteria" (Crotty 2003, p. 3). Contingency theory is the theoretical perspective of choice. "The essence of the contingency theory paradigm is that organizational effectiveness results from fitting characteristics of the organization [...] to contingencies that reflect the situation of the organization" (Donaldson 2001, p. 1)[17]. Contingencies include the environment, organizational size and strategy and organizational attempts to attain fit of organizational characteristics, such as organizational culture, to these contingencies because fit leads to high performance (cf. Donaldson 2001, p. 1 f.). Therefore, to avoid decreased performance organizations adopt new organizational characteristics that fit new levels of contingencies (cf. Donaldson 2001, p. 2). In other words, organizations attain fit of the organizational culture with contingencies to succeed in obtaining fit between the organizational characteristics and contingencies.

Finally, the epistemology, "the theory of knowledge embedded within the theoretical perspective and thereby in the methodology" (Crotty 2003, p. 3) must be described. "Epistemology is concerned with providing a philosophical grounding for deciding what kinds of knowledge are possible and how we can

---

[16] Author's own figure, referencing *Crotty* (2003, p. 2).
[17] *Donaldson* (2001), referencing *Burns & Stalker* (1961), *Lawrence & Lorsch* (1967), *Pennings* (1992) and *Woodward* (1995.)

ensure that they are both adequate and legitimate" (Maynard 2002, p. 10). From the range of possible epistemologies (objectivist, constructivist to a subjectivist epistemology), in this thesis objectivism is the epistemology of choice because it assumes that culture exists as a discrete entity. Culture is one of the elements of an organization, and so an organization has a culture, which has a purpose and function (cf. Janićijević 2011, p. 74). "Objectivist epistemology holds that meaning, and therefore meaningful reality, exists as such apart from the operation of any consciousness. […] In this objectivist view of 'what it means to know', understanding and values are considered to be objectified in the people we are studying and, if we go about in the right way, we can discover the objective truth" (Crotty 2003, p. 8).

Section 1.5 outlines a rough structure of this thesis. The structure of this thesis arises out of both the deduced research question and this thesis's formulated global objective and sub-objectives.

1.5 Structure

The first section, **'The Impact, Meaning and Challenges of Knowledge Absorption'**, begins with an overview of SMEs' external knowledge acquisition. Subsequently, the state of research on the absorptive capacity of SMEs and organizational culture is discussed, identifying the research gap in these fields and formulating the objective, research process and structure of this thesis.

To develop a model of SMEs' external knowledge absorption, the conceptual principles of absorptive capacity and of organizational culture are first discussed in the second section, **'The Conceptual Principles'**. With regard to the conceptual principles of absorptive capacity, it first is pointed out that absorptive capacity serves as an important performance-enhancing lever for SMEs. Next, the role of the knowledge source and complementarity and experience as the key antecedents of absorptive capacity are explained. Absorptive capacity consists of two subsets: 'potential absorptive capacity' and 'realized absorptive capacity'. The four capabilities of acquisition, assimilation, transformation and exploitation of the model of absorptive capacity, based on *Zahra & George* (2002), can be combined into those two subsets of absorptive capacity. These conceptual principles will be discussed in their entirety in section 2.1. With regard to the conceptual principles of organizational culture explained in section 2.2, the term 'organizational culture' first must be defined, to allow a detailed consideration of organizational culture. Next, various approaches to organizational culture are explained. Culture has been conceptualized in terms of a number of dimensions, but although culture has been studied through the use of a number of dimensions, several studies agree on the important functions of culture (e.g., Smircich 1983, Koberg & Chusmir 1987, Girdauskienė & Savanevičienė 2007, Bhagat et al. 2002, Glisby & Holden, Nonaka 1994, Nonaka & Takeuchi 1995, Girdauskienė & Savanevičienė 2007, Holden 2001, Lemken et al. 2000, Lucas 2006, Moffett et al. 2002, Weissenberger-Eibl & Spieth 2006). A knowledge-friendly approach to organizational culture will be explained in detail.

The third section, **'A Model of External Knowledge Absorption'**, develops a model of external knowledge absorption by SMEs. The modeling is carried out in three steps. In the first step, the theoretical framework is modeled and suitable theories for explaining the relationship between organizational culture and absorptive capacity are identified: organizational learning, innovation, dynamic capabilities, the knowledge-based view, managerial cognition and coevolution. Based on this framework, in the second step, the parameters of the model that are relevant to identifying the starting points of a cultural design concept for external knowledge absorption, and in particular, organizational culture contexts, are worked out: absorptive capacity and organizational culture. With regard to absorptive capacity, the four capabilities of acquisition, assimilation, transformation and exploitation are explained in detail. With regard to organizational culture, the six dimensions of trust, collaboration, openness, autonomy, learning receptivity and care are explained in detail. After explaining how the parameters of the model of the relationships among the dimensions of organizational culture and the capabilities of absorptive capacity are worked out, they are transferred into an empirically testable model in a third step.

The objective of the analysis in the fourth section, **'An Empirical Analysis of the Research Models'**, is to evaluate the model of SMEs' external knowledge absorption. To conduct the empirical analysis of the research model, the first step defines SMEs as the object of study. The second step explains the process of data collection. Hypotheses about the relationships among the dimensions of organizational culture and the capabilities of absorptive capacity that are represented in the model are tested via the quantitative research method of a survey, allowing for SMEs' organizational culture to be systematically captured. To capture the data, all of the variables are operationalized. Subsequently, the gathered data are analyzed and the results are discussed and summarized.

Section **'Summary, Conclusion and Outlook'**, concludes this thesis with an overview of its findings and contributions to and implications for management and research. The first section 5.1 begins with an overall summary of this thesis and its contribution to management and research. The second section 5.2 addresses implications for management with respect to trust, collaboration, openness and learning receptivity, which are the dimensions of organizational culture that are relevant for the support of absorptive capacity. The third section 5.3 discusses limitations of the investigation and further research needs.

Figure 2 graphically summarizes this thesis's structure and course of investigation.

1 The Impact, Meaning and Challenges of Knowledge Absorption

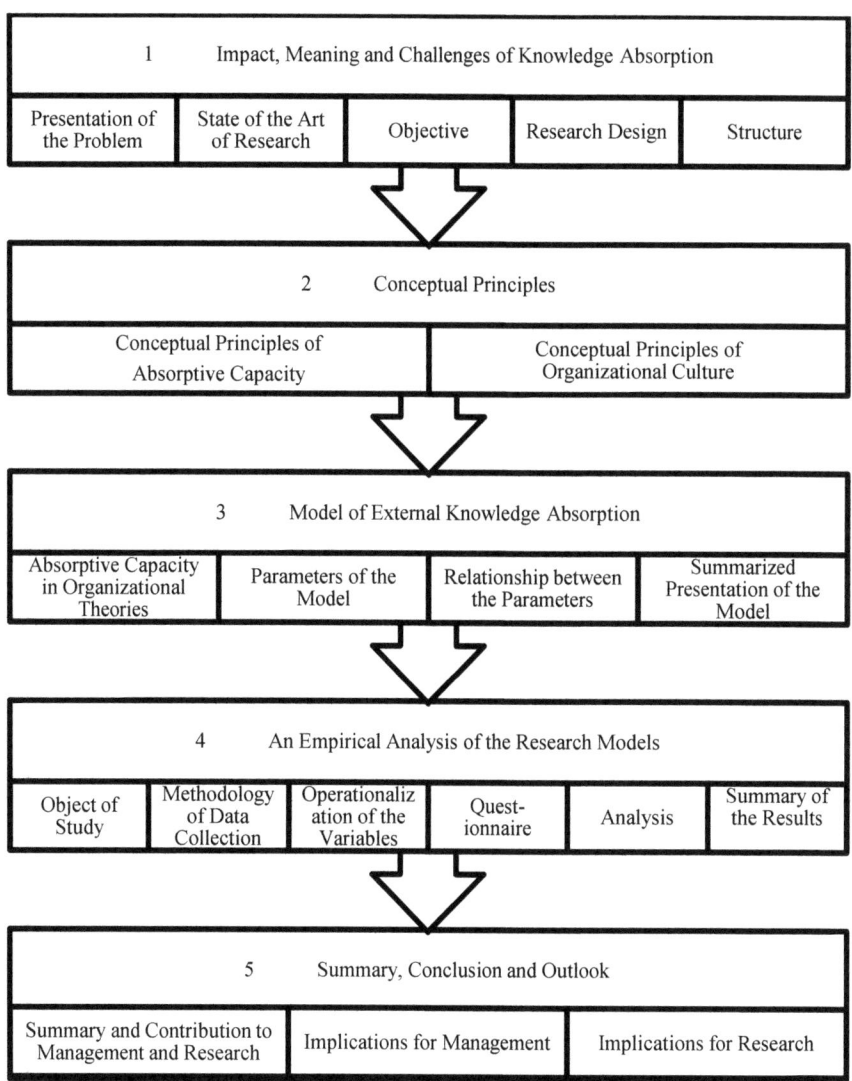

Figure 2 Structure of This Thesis[18]

---

[18] Author's own figure.

2 The Conceptual Principles

With regard to the conceptual principles of absorptive capacity, section 2.1.1 first points out how absorptive capacity serves as an important performance-enhancing layer for SMEs. Next, section 2.1.2 explains what role the knowledge source and complementarity and experience as the key antecedents of absorptive capacity play for the two subsets of absorptive capacity, namely 'potential absorptive capacity' and 'realized absorptive capacity', which are described in section 2.1.3. In the end, section 2.1.4 explains in more detail that these two subsets consist of the four capabilities — acquisition, assimilation, transformation and exploitation — of the model of absorptive capacity, based on *Zahra & George* (2002), before discussing the conceptual principles of organizational culture.

According *Schein*'s (1984 & 1990) definition, organizational culture manifests itself at three levels: observable artifacts, values and basic underlying assumptions. Section 2.2.1 explains these three levels of organizational culture. Next, section 2.2.2 explains approaches to organizational culture. Section 2.2.3 explains one special approach, a knowledge-friendly organizational culture approach of organizational culture, as the designated approach of this thesis.

2.1 The Conceptual Principles of Absorptive Capacity

Although *Kedia & Bhagat* (1988) first coined the term 'absorptive capacity' in their study of the cultural constraints on transfer of technology across nations, the contribution by *Cohen & Levinthal* (1990) is generally accepted as the founding paper on absorptive capacity (cf. Volberda et al. 2010, p. 932). *Cohen & Levinthal* (1989 & 1990) link a firms' R&D availability to its learning and innovative capabilities and therefore established a concept that is accepted as "a key concept for the literature on knowledge and innovation management of the past two decades" (Flatten, Greve & Brettel 2011b, p. 137)[19].

Cohen & Levinthal (1990) define absorptive capacity as "the ability of a firm to recognize the value of new, external information, assimilate it, and apply it to commercial ends" (Cohen & Levinthal 1990, p. 128). This definition highlights three steps of absorptive capacity, namely, knowledge recognition, knowledge assimilation and knowledge exploitation to commercial ends. Several re-conceptualizations of this original absorptive capacity construct by *Cohen & Levinthal* (1990) have appeared in the literature since 1990 (e.g., Lane et al. 2006, Todorova & Durisin 2007, Zahra & George 2002). For example, *Mowery & Oxley* (1995) propose that absorptive capacity "includes a broad array of skills, reflecting the need to deal with the tacit components of the transferred technology, as well as the frequent need to modify a foreign-sourced technology for domestic applications" (Mowery & Oxley 1995, S 81). *Zahra & George* (2002) analyze this and further definitions of absorptive capacity and concluded by referencing *Cohen & Levinthal* (1990), *Kim* (1998) and *Mowery & Oxley* (1995) for the proposition that absorptive capacity is "a multidimensional construct involving the ability to value, assimilate, and apply knowledge […] or is a combination of effort and knowledge bases" (Zahra & George 2002, p. 186). They highlight absorptive capacity "as a dynamic capability pertaining to knowledge creation and utilization that enhances a firm's ability to gain and sustain competitive advantage" (Zahra & George 2002, p. 185).

After reviewing the work on absorptive capacity in the past, *Zahra & George* (2002) reconceptualize original work by *Cohen & Levinthal* (1990) in developing their model of absorptive capacity (cf. Figure 3). They define absorptive capacity as "the set of organizational routines and processes by which firms acquire, assimilate, transform, and exploit knowledge to produce a dynamic capability" (Zahra & George 2002, p. 186). Their definition and model of

---

[19] The terms 'construct' and 'concept' are used synonymously throughout this thesis.

absorptive capacity is characterized by three changes in the key antecedents, moderators and outcomes of the construct, its components and its capabilities, which are described in sections 2.1.2 through 2.1.4.

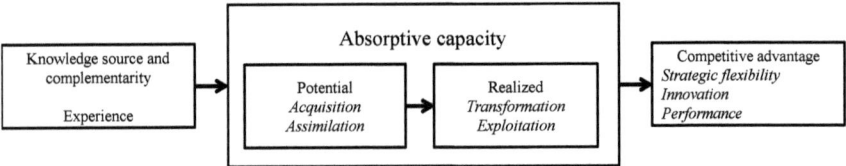

Figure 3 A Model of Absorptive Capacity[20]

The model of absorptive capacity by *Zahra & George* (2002) is used as the basis model for this thesis. Before the model is explained in more detail, the relationship among the capabilities of absorptive capacity, competitive advantage and firm performance in SMEs is discussed in the following section 2.1.1 because it has already been addressed several times that absorptive capacity serves as an important performance-enhancing lever for SMEs.

2.1.1 The Relationship between Absorptive Capacity and Firm Performance

Organizational resources and capabilities that lead to competitive advantage are both valuable and unlikely to be available from others. This statement has been supported by *Barney* (1991): "A firm is said to have a *sustained competitive advantage* when it is implementing a value creating strategy not simultaneously being implemented by any current or potential competitors *and* when these other

---

[20] Author's own figure, referencing *Zahra & George* (2002, p. 192); the moderators of absorptive capacity (activation triggers, social integration mechanisms and regimes of appropriability) are not depicted because they are not the focus of this thesis. They are described in detail in *Zahra & George* (2002, p. 192 f.).

firms are unable to duplicate the benefits of this strategy" (Barney 1991, p. 102). Absorptive capacity is a bundle of knowledge-based capabilities and can be a source of a firm's competitive advantage because differences in firms' utilization of organizational resources and capabilities impact performance (cf. Zahra & George 2002, p. 195).

According to the model of absorptive capacity by *Zahra & George* (2002), firms can achieve competitive advantage through innovation and strategic flexibility, and therefore, realized absorptive capacity influences a firm's performance through product and process innovation (cf. Zahra & George 2002, p. 195). Transformation capabilities help firms to develop new perceptual schema or changes to existing processes[21]. Exploitation capabilities help firms to convert knowledge into new products and services[22].

SMEs often have to deal with a lack of resources and capabilities that are required to complete those processes that generate long-term competitive advantages. As an example of external knowledge sources, strategic alliances present a good opportunity to overcome this deficiency through sharing costs and risks[23]. Due to their lack of resources and capabilities, SMEs typically cannot heavily invest in R&D activities, but engagement in strategic alliances is a way to develop and expand an SME's knowledge base without making a large investment (cf. Flatten et al. 2011, p. 139). *Flatten et al.* (2011b) find some multifaceted relationships between strategic alliances, absorptive capacity and firm performance: "Strategic alliances have a strong effect on firm performance, while ACAP influences both firm performance and success of strategic alliances" (Flatten et al. 2011b, p. 146)[24]. To provide further detail, their findings imply that firms with a barely developed absorptive capacity cannot handle external knowledge as successfully as firms with a highly developed absorptive ca-

---

[21] Transformation capabilities are explained in detail in section 3.2.1.3.
[22] Exploitation capabilities are explained in detail in section 3.2.1.4.
[23] External knowledge sources are explained in detail in section 2.1.2.
[24] 'ACAP' is an abbreviation of 'absorptive capacity'.

pacity and therefore, they differ with regard to their ability to enhance their performance by engaging in strategic alliances: "SMEs with a well-developed ACAP can additionally use it as an instrument to improve the effectiveness of strategic alliances, and thus enhance firm performance" (Flatten et al. 2011b, p. 147).

To summarize, absorptive capacity serves as an important performance-enhancing lever for SMEs, which typically have a lack of internal resources and therefore tend to depend heavily on absorptive capacity because external knowledge absorption is a critical factor for competitiveness.

The following sections explain the model of absorptive capacity by *Zahra & George* (2002). In their model, external sources of knowledge and experience are the key antecedents of absorptive capacity. Both antecedents are explained in the following section 2.1.2.

2.1.2 Antecedents of Absorptive Capacity

A firm's attempt to acquire knowledge from external sources is based on an idea the firm has for an application for a new product or service, but about which the firm lacks relevant knowledge (cf. Lichtenthaler & Lichtenthaler 2012, p. 161). External knowledge sources include acquisitions (e.g., Chaudhuri & Tabrizi 1999), communications (e.g., Levine, Higgins & Choi 2000, Stasser, Vaughan & Stewart 2000, Warkentin & Beranek 1999), interactions with suppliers and customers (e.g., von Hippel 1988), interorganizational relationships, including R&D consortia, strategic alliances and joint ventures (e.g., Baum & Ingram 1998, Darr, Argote & Epple 1995, Das & Teng 1998, Ding, et al. 2009, Lane & Lubatkin 1998, Larsson, Bengtsson, Henriksson & Sparks 1998, McEvily & Zaheer 1999, Powell et al. 1996, Simonin 1999, Vermeulen & Barkema 2001), observations (e.g., Nonaka, Umemoto & Senoo 1996), patents (e.g., Appleyard 1996), purchasing through licensing and contractual agreements (e.g., Granstrand & Sjölander 1990), staff transfers (e.g., Almeida & Kogout 1999,

Gruenfeld, Martorana & Elliott 2000) and training measures (e.g., Moreland & Myaskovsky 2000, Thompson et al. 2000). *Maurer & Tiwana* (2012) highlight that the efficient integration of knowledge from multiple sources results in greater performance than relying solely on internal knowledge (cf. Maurer & Tiwana 2012, p. 4). These multiple sources are summarized in Table 1.

Table 1 External Knowledge Sources[25]

Source	Author
Acquisitions	e.g., *Chaudhuri & Tabrizi* (1999)
Communications	e.g., *Levine et al.* (2000), *Stasser, Vaughan & Stewart* (2000), *Warkentin & Beranek* (1999)
Interactions with suppliers and customers	e.g., *von Hippel* (1988)
Interorganizational relationships	e.g., *Baum & Ingram* (1998), *Darr et al.* (1995), *Das & Teng* (1998), *Ding et al.* (2009), *Lane & Lubatkin* (1998), *Larsson et al.* (1998), *McEvily & Zaheer* (1999), *Powell et al.* (1996), *Simonin* (1999), *Vermeulen & Barkema* (2001)
Observations	e.g., *Nonaka et al.* (1996)
Patents	e.g., *Appleyard* (1996)
Purchasing through licensing and contractual agreements	e.g., *Granstrand & Sjölander* (1990)
Staff transfers	e.g., *Almeida & Kogout* (1999), *Gruenfeld et al.* (2000)
Training measures	e.g., *Moreland & Myaskovsky* (2000), *Thompson et al.* (2000)

Firms can acquire knowledge from different external knowledge sources, and the diversity of these sources significantly influences the acquisition and assimilation capabilities and therefore, a firm's potential absorptive capacity (cf. Zahra & George 2002, p. 191 f.).

---

[25] Author's own table.

To absorb external knowledge that will enhance a firm's stock of knowledge to achieve competitive advantage, a firm needs absorptive capacity. The ability to absorb knowledge depends on a firm's experience because this ability is largely a function of the level of prior related knowledge (cf. Cohen & Levinthal 1990, p. 129). *Rosenkopf & Nerkar* (2001) point out that a firm's assimilation capability is strongly associated with its past R&D activity (cf. Rosenkopf & Nerkar 2001, p. 287). It is important that a firm's local search is affected by past experience because its R&D activities are closely related to previous R&D activities (cf. Rosenkopf & Nerkar 2001, p. 287). Directing areas of a local knowledge search to past experience in which the firm has had past success influences the development of future acquisition capabilities (cf. Zahra & George 2002, p. 193).

In addition to experience, a further key for the absorption of knowledge is path dependency (cf. Cohen & Levinthal 1990, p. 135). *Zahra & George* (2002) highlight the role of the path dependency of absorptive capacity because relevant prior knowledge forms the content of a firm's absorptive capacity (cf. Zahra & George 2002, p. 191). In the literature, the path dependency of absorptive capacity is widely acknowledged by researchers such as *Cohen & Levinthal* (1990) and *Lane et al.* (2006), as *Lichtenthaler & Lichtenthaler* (2012) highlight because prior knowledge in a particular field determines a firm's ability to profit from external knowledge in that field (cf. Lichtenthaler & Lichtenthaler 2012, p. 162). Prior knowledge permits the absorption of new knowledge (cf. Cohen & Levinthal 1990, p. 135 f.). Therefore, it is important that some prior knowledge be very closely related to the new knowledge to facilitate the absorption of external knowledge. This can be explained as follows: "Accumulating absorptive capacity in one period will permit its more efficient accumulation in the next. By having already developed some absorptive capacity in a particular area, a firm may more readily accumulate what additional knowledge it needs in the subsequent periods in order to exploit any critical external knowledge that may become available" (Cohen & Levinthal 1990, p. 135 f.).

2 The Conceptual Principles

In summary, a firm's prior knowledge enables it to add value to incoming factors of production, namely external knowledge, in a unique manner that allows the firm to gain competitive advantage (cf. Spender 1996, p. 45). *Spender* (1996) points this out as follows: "So long as we assume markets are reasonably efficient and that competitive advantage is not wholly the consequence of asymmetric information about those markets or the stupidity of others, these rent-yielding capabilities must originate within the firm if they are to be of value. Since the origin of all tangible resources lies outside the firm, it follows that competitive advantage is more likely to arise from the intangible firm-specific knowledge which enables it to add value to the incoming factors of production in a relatively unique manner" (Spender 1996, p. 45).

In section 2.1.3, the components of absorptive capacity, which are essential to developing and increasing the knowledge base of a firm, are explained.

2.1.3 Components of Absorptive Capacity

Zahra & George (2002) ascertain that absorptive capacity consists of two components: potential absorptive capacity and realized absorptive capacity (cf. Zahra & George 2002, p. 190). These two components of absorptive capacity have different value-creating potentials: Potential capacity centers on the acquisition and assimilation capability of knowledge and realized capacity centers on the transformation and exploitation capability of knowledge (cf. Zahra & George 2002, p. 185). Furthermore, potential absorptive capacity and realized absorptive capacity have different but complementary roles by coexisting at all times and fulfilling a necessary but insufficient condition to improve a firm's competitive advantage (cf. Zahra & George 2002, p. 190). *Zahra & George* (2002) give the example that on the one hand, firms cannot possibly exploit knowledge without first acquiring it, but on the other hand, the capability to transform and exploit knowledge for profit generation might be missing after successful knowledge acquisition and assimilation, so that high potential absorptive capacity does not necessarily imply enhanced performance (cf. Zahra & George 2002, p. 190).

The theoretical distinction between potential and realized absorptive capacity is important for three reasons related to the evaluation of the contributions for a firm's competitive advantage[26]:

1. It can be explained through inefficiency in leveraging potential absorptive capacity and therefore, missing performance improvements, why certain firms are more efficient than others in using absorptive capacity. Therefore, it is possible to compare the different contributions of the potential and realized absorptive capacity to building the firm's competitive advantage.
2. Different managerial roles that are necessary to nurture and harvest the two components of absorptive capacity can be examined.
3. The distinction between potential and realized absorptive capacity provides a basis for observing and examining the fluid and nonlinear paths that organizations may follow in developing their core competencies, and it is possible to study why some firms fail because of changes in their external environments.

Section 2.1.4 explains the several capabilities of potential and realized absorptive capacity.

2.1.4 Capabilities of Absorptive Capacity

In comparison to *Cohen & Levinthal* (1989), who define absorptive capacity as an ability that facilitates the accumulation and subsequent use of knowledge, *Zahra & George* (2002) broaden the concept of absorptive capacity, because the exploitation of externally acquired knowledge usually requires the conversion of

---

[26] Cf. *Zahra & George* (2002, p. 190), extends to the listing.

2 The Conceptual Principles

its content into a usable form (cf. Flatten et al. 2011a, p. 98). They broaden this concept from the original three capabilities of *Cohen & Levinthal* (1989), namely identification, assimilation, and exploitation, to four capabilities, namely acquisition and assimilation, both building the potential absorptive capacity, and transformation and exploitation, both building the potential realized capacity (cf. Figure 4).

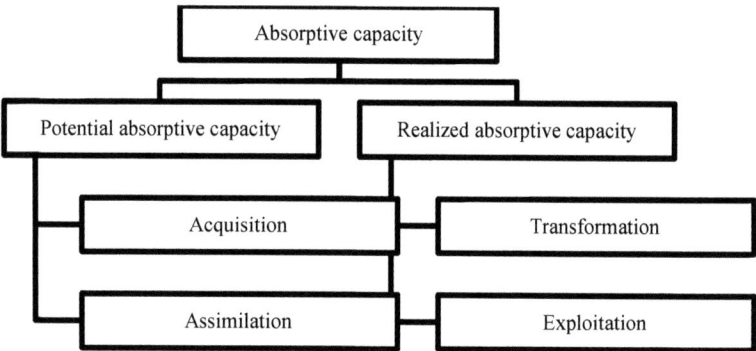

Figure 4 Capabilities of Absorptive Capacity[27]

The four capabilities can be described as follows[28]:

- The **acquisition capability** refers to a firm's capability to identify and acquire knowledge relevant to a company's operations from external knowledge sources (cf. Flatten et al. 2011a, p. 100, cf. Flatten et al. 2011b, p. 138, cf. Zahra & George 2002, p. 189). Prior knowledge is considered as a prerequisite in this process that positively influences

---

[27] Author's own figure.
[28] A detailed description of the four capabilities of absorptive capacity follows in section 3.2.1, which describes the parameters of the model. Cf. the description of the capabilities of absorptive capacity in *Ivens, Zerwas & Schaarschmidt* (2014).

the absorption of external knowledge (cf. Cohen & Levinthal 1990, p. 128).

- The **assimilation capability** refers to a firm's capability to develop routines and processes that allow analyzing, processing, interpreting, and understanding of external sources of acquired knowledge (cf. Flatten et al. 2011a, p. 100[29], cf. Zahra & George 2002, p. 189). It is important for firms to develop new routines because those routines influence the locus of the future search for knowledge (cf. Zahra & George 2002, p. 193).
- The **transformation capability** refers to a firm's capability to develop and refine those routines that facilitate the combining of existing knowledge with the acquired and assimilated knowledge for future use (cf. Flatten et al. 2011a. p. 100, cf. Zahra & George 2002, p. 190).
- The **exploitation capability** refers to a firm's capability to refine, extend, and leverage existing routines, competencies and technologies or to create new ones by incorporating acquired and transformed knowledge into a firm's own operations (cf. Flatten et al. 2011a, p. 100[30], cf. Zahra & George 2002, p. 190).

The four capabilities are distinct but complementary (cf. Zahra & George 2002, p. 189).

Together, the four capabilities of acquisition, assimilation, transformation and exploitation that compose a firm's absorptive capacity enable companies to exploit new technological developments (cf. Cohen & Levinthal 1994, p. 227). *Eisenhardt & Martin* (2000) state that the capabilities have equifinality, homogeneity and substitutability across firms (cf. Eisenhardt & Jeffrey 2000, p. 1108). Although the capabilities are certainly idiosyncratic in their details, the capabilities also exhibit common features that are associated with effective

---

[29] *Flatten et al.* (2011a, p. 189 f.), referencing Szulanski (1996).
[30] *Flatten et al.* (2011a, p. 190), referencing *Del Carmen Haro-Domínguez, Arias-Aranda, Javier Lloréns-Montes & Ruíz Moreno* (2007).

processes across different firms (cf. Eisenhardt & Jeffrey 2000, p. 1108). Following *Eisenhardt & Martin* (2000), *Zahra & George* (2002) summarize that the "capabilities have some commonalities across different firms and attain equifinality, they are idiosyncratic in the specific ways firms pursue, develop, and employ them" (Zahra & George 2002, p. 189). They highlight that this is the crucial variability that gives firms a basis for developing different types of competitive advantage (cf. Zahra & George 2002, p. 189). Absorptive capacity achieves competitive advantage primarily through strategic flexibility, innovation and performance innovation (cf. Zahra & George 2002, p. 195 f.).

2.2 The Conceptual Principles of Organizational Culture

With regard to the conceptual principles of organizational culture, the term 'organizational culture' must first be defined to allow for a detailed consideration of organizational culture in section 2.2.1. Afterwards, in section 2.2.2, approaches to organizational culture are explained. In the literature, there are a variety of approaches to explain and operationalize organizational culture. This thesis is unable to conduct a full exposition of this widespread research field. Rather, it develops a fundamental understanding of the phenomenon of corporate culture that is sufficiently viable to take the following steps with regard to the objective of this thesis. In conclusion, in section 2.2.3, a knowledge-friendly organizational culture approach of organizational culture is explained as this thesis's designated approach.

2.2.1 The Elements and Levels of Organizational Culture

Organizational culture is a "continuous re-creation of 'shared meanings'" (Roskin 1986, p. 4) and is typically represented by the model of organizational culture of *Schein* (1984 & 1990) (cf. Lim 1995, p. 16). Schein (1984) defines organizational culture as "the *pattern of basic assumptions* that a *given group* has *invented, discovered, or developed in learning to cope* with its *problems of external adaption and internal integration* and that have *worked well enough to be considered valid*, and, therefore, to be *taught to new members* as the correct way to *perceive, think, and feel* in relation to those problems" (Schein 1984, p. 3). He understands organizational culture as a pattern of solutions to problems that have worked well in the past and are therefore taught to new members (cf. Schein 1984, p. 3). *Lemken et al.* (2000) support this understanding by viewing organizational culture as "the manner in which an organization solves problems to achieve its specific goals and to maintain itself over time" (Lemken et al. 2000, p. 3).

Based on his definition of organizational culture, *Schein* (1990) concludes that the strength and degree of the internal consistency of an organizational culture are "a function of the stability of the group, the length of time the group has existed, the intensity of the group's experiences of learning, the mechanisms by which the learning has taken place (i.e., positive reinforcement or avoidance conditioning), and the strength and clarity of the assumptions held by the founders and leaders of the group" (Schein 1990, p. 111). In analyzing organizational culture, it is possible to distinguish among three fundamental levels at which organizational culture manifests itself: observable artifacts, values and basic underlying assumptions (cf. Schein 1984, p. 4):

Artifacts and creation: Observable artifacts include "everything from the physical layout, the dress code, the manner in which people address each other, the smell and feel of the place, its emotional intensity, and other phenomena, to the more permanent archival manifestations such as company records, products, statements of philosophy, and annual reports" (Schein 1990, p. 111). *Schein* (1990) illustrates observable artifacts using the following example: "When one enters an organization one observes and feels its *artifacts*" (Schein 1990, p. 111). Therefore, artifacts are the material concretization of the organizational culture and help people to orient in and deal with the environment (cf. Spieth 2006, p. 71). Analysis of organizational culture at this level is tricky because the data are easy to obtain but hard to interpret: "We can describe 'how' a group constructs its environment and 'what' behavior patterns are discernible among the members, but we often cannot understand the underlying logic — 'why' a group behaves the way it does" (Schein 1984, p. 3). There is also a problem with observable artifacts that *Schein* (1990) highlights: "The problem with artifacts is that they are palpable but hard to decipher accurately. We know how we react to them, but that is not necessarily a reliable indicator of how members of the organization react. We can see and feel that one company is much more formal and bureaucratic than another, but that does not tell us anything about why this is so or what meaning it has to the members" (Schein 1990, p. 111 f.).

Values: To analyze 'why' a group behaves the way it does, the values that govern behavior, which is the second level in Figure 5, must be analyzed. Values can be differentiated between espoused and documented values and explain why, in addition to a firm's norms, ideologies, charters and philosophies, certain observed phenomena occur in the way they do (cf. Schein 1990, p. 111). The values of an organizational culture are hard to observe directly, but can be studied through interviews, questionnaires and survey instruments (cf. Schein 1990, p. 111). Although such values can be identified and studied, they represent only the manifest or espoused values of the organizational culture: "That is they focus on what people say in the reason for their behavior, what they ideally would like those reasons to be, and what are often their rationalizations for their behavior" (Schein 1984, p. 3).

Assumptions: *Schein* (1984) draws inferences from his research in 1981 and 1983 about the opportunity to understand the organizational culture and to ascertain more completely the values and overt behavior of an organization through cultural analysis: "Such analysis is, of course, common when we think of ethic or national cultures, but not sufficient attention has been paid to the possibility that groups and organizations within a society also develop cultures that affect in a major way how the members think, feel, and act" (Schein 1985, p. 3; own formatting). Organizational culture manifests itself at the level of assumptions, "taken-for-granted, underlying, and usually unconscious *assumptions* that determine perceptions, thought processes, feelings, and behavior" (Schein 1990, p. 111). *Schein* (1984) explains the taken-for-granted, underlying, and usually unconscious character of assumptions as following: "Such assumptions are themselves learned responses that originated as espoused values. But, as a value leads to a behavior, and as that behavior begins to solve the problem which prompted it in the first place, the value gradually is transformed into an underlying assumption about how things really are. As the assumption is increasingly taken for granted, it drops out of awareness" (Schein 1984, p. 3 f.). *Schein* (1990) points out that understanding assumptions helps one to understand the meanings implicit in various observed behaviors (cf. Schein 1990, p. 111). In comparison to values, assumptions can be studied through more intensive observation and focused questions and by involving motivated members of a group in intensive self-analysis (cf. Schein 1990, p. 111).

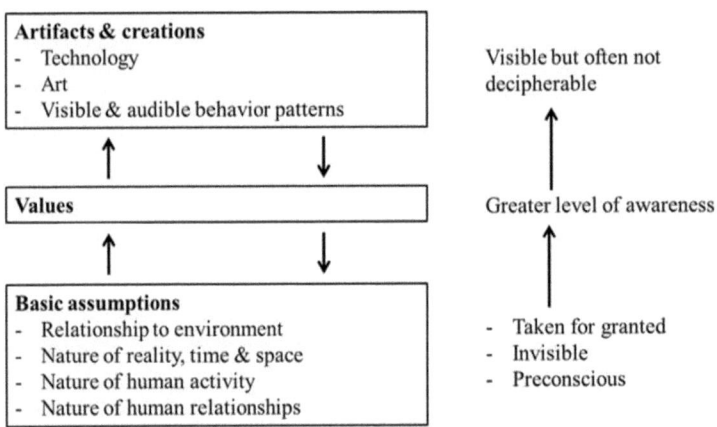

Figure 5 Levels of Culture and Their Interaction[31]

---

[31] Author's own figure, referencing *Schein* (1984, p. 4).

For a deeper analysis of organizational culture, a detailed conceptualization of the construct must be undertaken. Since the late 1970s, there have been several quantitative approaches, which often are based on the approach of *Schein* (1985), which address the measurement and operationalization of the construct of organizational culture. Section 2.2.2 describes several approaches to organizational culture.

2.2.2 Approaches to Organizational Culture

Culture has been conceptualized in terms of a number of dimensions, but although many approaches to organizational culture exist in literature, relevant authors have coined the term 'organizational culture' and developed approaches that have been used repeatedly in the research, such as the approach of *Gordon & Cummins* (1979), *Hofstede* (1980), *Reynolds* (1986), *O'Reilly et al.* (1991), *Hofstede* (1998) and others. For example, *Gordon & Cummins* (1979) classify organizational culture along the dimensions of 'organizational clarity', 'decision-making structure', 'organizational integration', 'management style', 'performance orientation', 'organizational vitality', 'compensation' and 'human resource development', whereas *Hofstede* (1980) consider organizational culture along the dimensions of 'power distance', 'uncertainty avoidance', 'individualism' and 'masculinity' (cf. Hofstede 1980). *Hofstede*'s (1980) dimensions are ubiquitous in the culture-related literature, and the multitude of citations and replications of Hofstede's dimensions suggest the significant impact of his work on culture research since 1980 (cf. King 2007). Although each of *Hofstede*'s (1980) dimensions has been supported and criticized by several authors, collectivism and individualism in particular have been identified by various authors as essential dimensions for understanding differences in organizational culture (e.g., Bhagat et al. 2002, Chen, Chen & Meindl 1998, Anakwe, Kessler & Christensen 1999, Azevedo, Drost & Mullen 2002)[32]. Examples of approaches that do not address collectivism and individualism are, e.g., *Cooke & Laferty* (1989) and *O'Reilly et al.* (1991). *Cooke & Laferty* (1989) classifies organizational culture along the dimensions of 'humanistic/helpful', 'affiliation', 'achievement', 'self-actualization', 'approval', 'conventionality', 'dependence', 'avoidance', 'oppositional', 'power', 'competitive' and 'perfectionism', whereas *O'Reilly et al.* (1991) instead choose to view this phenomenon through the dimensions 'innovation', 'outcome orientation', 'aggressiveness', 'detail orientation', 'stability' and 'respect for people'.

These examples show that culture has been conceptualized in terms of a number of dimensions, but although culture has been studied through the use of a number of dimensions, several studies agree on culture's important functions (e.g., Smircich 1983, Koberg & Chusmir 1987, Girdauskienė & Savanevičienė 2007, Bhagat et al. 2002, Glisby & Holden, Nonaka 1994, Nonaka & Takeuchi 1995, Girdauskienė & Savanevičienė 2007, Holden 2001, Lemken et al. 2000, Lucas 2006, Moffett et al. 2002, Weissenberger-Eibl & Spieth 2006). The following tables (cf. Table 2 through Table 6) give an overview of the conceptualization and operationalization of organizational culture by several authors[33]: *Gordon & Cummins* (1979), *Allen & Dyer* (1980), *Hofstede* (1980),

---

[32] In a subsequent investigation, *Hofstede* (1998) classifies organizational culture along the dimensions of 'communication climate', 'attitudes about work content', 'values about work context', 'gender issues', 'attitudes about direct boss', 'attitudes towards work pressures' and 'values about work content' (cf. Hofstede 1998).

[33] This thesis compares several approaches to find the most suitable approach of organizational culture for this investigation. *Spieth* (2009) and *Unterreitmeier* (2004) give a comprehensive overview of several approaches to organizational culture and compare them with regard to the concept of corporate culture, design of the measurement instrument, dimensions, indicator, scale, sample shape and type of data analysis (cf. Spieth 2006, p. 186 ff., cf. Unterreitmeier 2004, p. 86 ff.). The conceptualizations and operationalizations listed by *Spieth* (2009) and *Unterreitmeier* (2004) form the basis for comparing the qualifications of the approaches for this thesis, inter alia because *Spieth* (2009) discusses cultural constraints in transferring knowledge and therefore works in the knowledge management context, and because the objective of *Un-*

2 The Conceptual Principles

Reynolds (1986), *Cooke & Laferty* (1989), *Hofstede et al.* (1990), *Kern* (1991), *O'Reilly et al.* (1991), *Chatterjee et al.* (1992), *Fletcher & Jones* (1992), *Gordon & DiTomaso* (1992), *Denison & Mishra* (1995), *Weber* (1996), *Xenikou & Furnham* (1996), *Hofstede* (1998), *Poech* (2003), *Unterreitmeier* (2004) and *Sollberger* (2006)[34]. The conceptualizations and operationalizations of these authors are compared with regard to the concept of organizational culture, its dimensions and its indicators.

---

[34] *terreitmeier* (2004) was to develop an approach to measure organizational culture in terms of the best possible reliability and validity.

Spieth (2009) and *Unterreitmeier* (2004) both compare the approaches to the conceptualization and operationalization of organizational culture by *Gordon & Cummins* (1979), *Allen & Dyer* (1980), *Hofstede* (1980), *Glaser* (1983), *Kilmann & Saxton* (1983), *Sashkin* (1984), *Kobi & Wüthrich* (1986), *Reynolds* (1986), *Cooke & Laferty* (1988), *Hofstede, Neuijen, Ohayv & Sanders* (1990), *Kern* (1991), *O'Reilly, Chatman & Caldwell* (1991), *Chatterjee, Lubatkin, Schweiger & Weber* (1992), *Flechter & Jones* (1992), *Gordon & Di Tomaso* (1992), *Guptare* (1992), *Denison & Mishra* (1995), *Holmes & Mardsen* (1996), *Weber* (1996), *Xenikou & Furnham* (1996), *Hofstede* (1998) and *Poech* (2003) (cf. Spieth 2006, p. 186 ff., cf. Unterreitmeier 2004, p. 86 ff.). Furthermore, *Spieth* (2009) investigates the conceptualization and operationalization of organizational culture by *Unterreitmeier* (2004) and *Sollberger* (2006) (cf. Spieth 2006, p. 193). *Glaser* (1983), *Kilmann & Saxton* (1983), *Sashkin* (1984), *Kobi & Wüthrich* (1986), *Guptare* (1992) and *Holmes & Mardsen* (1996) are not considered within this thesis because they have such minor hit rates that the quality of the approaches is questionable.

Table 2 Approaches to Organizational Culture (I)[35]

Author(s)	Concept Of Organizational Culture	Design Of The Measurement Instrument	Dimensions	Indicator
Gordon & Cummins (1979)	Not explicitly defined because it is a measurement of the "Management Climate"	Qualitative interviews, analysis of documents, literature research (not specified), factor and cluster analysis	Organizational clarity; decision-making structure; organizational integration; management style; performance orientation (extract)	48 items
Allen & Dyer (1980)	Unconscious behavioral norms	Not specified	Performance facilitation; job involvement; training; leader-subordinate interaction; policies and procedures; confrontation; supportive climate	38 items
Hofstede (1980)	Not explicitly defined because it is a measurement of national cultures by means of working environment, values and work objectives	Expert discussions, preliminary studies, factor analysis	Power distance; uncertainty avoidance; individualism; masculinity	Initially approximately 80 items, proposed questionnaire for prospective intercultural studies contains 25 items.

---

[35] Author's own Table 2, referencing *Spieth* (2009), p. 186 ff. and *Unterreitmeier* (2004), p. 86 ff..

2 The Conceptual Principles

Table 3 Approaches to Organizational Culture (II)[36]

Author(s)	Concept Of Organizational Culture	Design Of The Measurement Instrument	Dimensions	Indicator
Reynolds (1986)	Sociostructural system (structures, strategies, management processes etc.); cultural system (values, beliefs)	Taken from literature (among others from *Deal & Kennedy* 1982, *Hofstede* 1980, *Harrison* 1972, *Peters & Waterman* 1982), subsequent reduction	External vs. internal emphasis; task vs. social focus; safety vs. risk; conformity vs. individuality; individual vs. group rewards; individual vs. collective; (extract)	14 items (5 response categories each)
Cooke & Laferty (1989)	Norms	Not specified	Humanistic/helpful; affiliation; achievement; self-actualization; approval; conventionality; dependence; avoidance; oppositional; power; competitive; perfectionism	120 items (10 for each dimension)
Hofstede et al. (1990)	Values and practices (symbols, heroes, rituals)	60 items taken from *Hofstede* (1980), 75 more items obtained by means of in-depth interviews and own reflections	Values (e.g., need for security; work centrality; need for authority); practices (e.g., process-oriented vs. results-oriented; employee- vs. job-oriented; parochial vs. professional)	135 items, reduced to 81 items

---

[36] Author's own Table 3, referencing *Spieth* (2009), p. 186 ff. and *Unterreitmeier* (2004), p. 86 ff..

Table 4 Approaches to Organizational Culture (III)[37]

Author(s)	Concept Of Organizational Culture	Design Of The Measurement Instrument	Dimensions	Indicator
Kern (1991)	Value maintenance and immediately observable expressions	Exploration (consultation with experts), factor analysis, item selection (subjective)	Humanity; international business policy; competitive orientation; environmental protection/safety; market/services; power/expansion; (extract)	143 items (reduced to 33)
O'Reilly et al. (1991)	Organizational values	54 statements (of unknown origin) condensed by means of factor analysis, further reduced to 26 items	Innovation; outcome orientation; aggressiveness; detail orientation; stability; respect for people	26 items
Chatterjee et al. (1992)	Not explicitly defined	Partly based on Gordon & Cummins (1979), Kilmann & Saxton (1983) and others	Innovation and action orientation; risk-taking; lateral integration; top management contact; autonomy and decision making; performance orientation; reward orientation	29 items
Fletcher & Jones (1992)	Underlying values, beliefs and principles, expressed in management structure and practices	Not specified	Work demands; interpersonal relationships in the workplace; work supports and constraints; physical environment; performance; organizational commitment; job dissatisfaction; strain	Not specified

---

[37] Author's own Table 4, referencing *Spieth* (2009), p. 186 ff. and *Unterreitmeier* (2004), p. 86 ff..

2 The Conceptual Principles

Table 5 Approaches to Organizational Culture (IV)[38]

Author(s)	Concept Of Organizational Culture	Design Of The Measurement Instrument	Dimensions	Indicator
Gordon & DiTomaso (1992)	Beliefs and values	Indicators according to Gordon & Cummins (1979), factor analysis	Clarity of strategy/shared goals; systematic decision making; integration/communication; innovation/risk-taking; (extract)	48 items
Denison & Mishra (1995)	Not specified	Own reflections	Adaptability; mission; involvement; consistency	8 items (2 for each dimension)
Weber (1996)	Values and basic assumptions according to Schein (1985)	Approach according to Chatterjee et al. (1992)	Innovation and action orientation; risk-taking; lateral integration; top management contact; (extract)	29 items
Xenikou & Furnham (1996)	Values, standards, artifacts	Concepts by Glaser (1983), Kilmann & Saxton (1983), Sashkin (1984), Cooke & Lafferty (1989)	Openness to change in a cooperative culture; task-oriented organizational growth; the human factor in a bureaucratic culture; artifacts; (extract)	218 items (all of the indicators of the consulted concepts)

---

[38] Author's own Table 5, referencing *Spieth* (2009), p. 186 ff. and *Unterreitmeier* (2004), p. 86 ff..

Table 6 Approaches to Organizational Culture (V)[39]

Author(s)	Concept Of Organizational Culture	Design Of The Measurement Instrument	Dimensions	Indicator
Hofstede (1998)	Attitudes, values and practices	Hofstede et al. (1990), expert interviews, item selection	Communication climate; attitudes about work content; values about work context; gender issues; attitudes about direct boss; attitudes toward work pressures; vales about work content	110 items (reduced to 73)
Poech (2003)	Center of Excellence-Model according to Frey & Schulz-Hardt (2000)	Preliminary study on item generation	Corporate culture; service culture; team culture; learning culture; innovation culture; management culture	60 items (10 for each dimension)
Unterreitmeier (2004)	Values and basic assumptions according to Schein (1985)	Meta-analysis of existing approaches and consultations with experts	Decision-making and leadership style; result and career orientation; employee orientation; remuneration justice; (extract)	120 items
Sollberger (2006)	Values and basic assumptions according to Schein (1985)	Preliminary study on item generation and meta-analysis	Trust; cooperation; openness; autonomy; willingness to learn; care	32 items

---

[39] Author's own Table 6, referencing *Spieth* (2009), p. 186 ff. and *Unterreitmeier* (2004), p. 86 ff..

2 The Conceptual Principles   49

The approaches presented in the previous tables mostly focus on partial aspects of the organizational culture and are based on different concepts of corporate culture, different designs of the measurement instrument, different dimensions and different indicators:

- **Concept of organizational culture**: With regard to the concept of corporate culture, *Hofstede* (1980), *Reynolds* (1986), *Hofstede et al.* (1990), *Kern* (1991), *O'Reilly et al.* (1991), *Fletcher & Jones* (1992), *Gordon & DiTomaso* (1992), *Weber* (1996), *Unterreitmeier* (2004) and *Sollberger* (2006) are devoted to values. In contrast, *Allen & Dyer* (1980) and *Cooke & Laferty* (1989), consider norms. Several of the approaches make combinations of the capturing of partial aspects such as values and norms, e.g., *Xenikou & Furnham* (1996) and *Hofstede* (1998).
- **Design of the measurement instrument**: For the design of the measuring instrument very different methods are used. The methods range from qualitative interviews, factor analysis, and personal reflections, to expert discussions and meta-analyses of existing approaches.
- **Dimensions and indicators**: The number of dimensions and indicators used is very heterogeneous: Some constructs of organizational culture consist of fewer than 5 dimensions, e.g., *Hofstede* (1980) and *Denison & Mishra* (1995). Other constructs of organizational culture consist of 10 or more dimensions, e.g., *Reynolds* (1986) and *Kern* (1991). To comport with the items for the several dimensions, the comparison shows that some studies use fewer than 30 items to measure organizational culture, e.g., *Reynolds* (1986) and *O'Reilly et al.* (1991). Other studies use more than 120 items, e.g., *Hofstede et al.* (1990) and *Kern* (1991), or even more than 200 items, e.g., *Xenikou & Furnham* (1996). The authors' scale, sample shape and type of data analysis all differed, too.

Because the literature has identified a generally positive influence of organizational culture on absorptive capacity with due regard to the theoretical concepts at the beginning of this thesis, a positive relationship between the dimensions of organizational culture and the capabilities of absorptive capacity is assumed[40]. With regard to the context of this thesis, *Sollberger* (2006) is the only one of the listed authors in the previous section who has worked out those dimensions of organizational culture that play a central role in the absorption of external knowledge and therefore are explained in more detail in section 2.2.3[41]. Her concept of organizational culture corresponds to the values and basic premises by *Schein* (1985) that is explained in section 2.2.1.

2.2.3 A Knowledge-Friendly Approach to Organizational Culture

Davenport, De Long & Beer (1997) introduce the 'knowledge-friendly culture' construct and claim that a knowledge-friendly culture is one of the most important factors that contribute to the success of the absorption of external knowledge (cf. Davenport et al. 1997, p. 14 f.). The reason for this claim is that it is perhaps the most difficult constraint that companies must address. These difficulties can be caused by several components relevant to a knowledge-friendly culture: a positive orientation to knowledge, absence of knowledge inhibitors in the culture and fit of the knowledge management project type with the existing culture (cf. Davenport et al. 1997, p. 15). Figure 6 shows examples of requirements that result from the three components.

---

[40] Although is the literature has distinguished between open and closed organizational cultures, investigations are usually based on an open and positive understanding of organizational culture. Therefore, in this thesis, given a spectrum of a closed to an open organizational culture, an open organizational culture is assumed.
[41] In this thesis the private sector is investigated and therefore the construct of organizational culture is used in another context than in the work of *Sollberger* (2006).

2 The Conceptual Principles

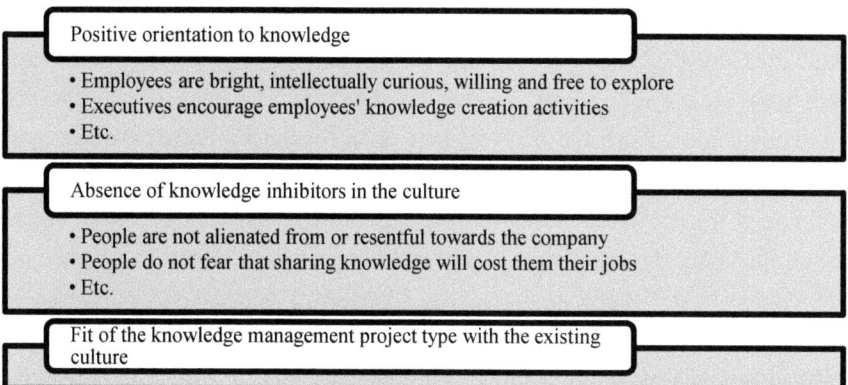

Figure 6 Components of a Knowledge-Friendly Organizational Culture[42]

Following this idea of a knowledge-friendly organizational culture, *Sollberger* (2006) emphasizes the difference between the organizational culture of a knowledge-oriented company — she calls this type of culture 'knowledge culture' — and the organizational culture of a traditional company, and states that knowledge culture is a part of organizational culture (cf. Sollberger 2006, p. 115 ff.)[43]. According to *Sollberger* (2006), knowledge culture "comprises the entirety of the norms and values of a company that shape the thinking and behavior of corporate members in the daily management of knowledge" (Sollberger 2006, p. 119)[44].

To define a knowledge culture in detail, *Sollberger* (2006) works out the dimensions of the organizational culture that play a central role in the inclusion and integration of a holistic knowledge management, and therefore the absorption of external knowledge, with the help of a detailed literature review

---

[42] Author's own figure, referencing *Davenport et al.* (1997, p. 15).
[43] The terms 'knowledge culture' and 'organizational culture' are used synonymously throughout this thesis, because a knowledge-friendly organizational culture is its focus.
[44] Translated by the author.

(cf. Sollberger 2006, p. 115 ff.). The results of the analysis by *Sollberger* (2006) show that 'trust', 'collaboration', 'openness', 'autonomy', 'learning receptivity' and 'care' are the dimensions of a knowledge culture (cf. Figure 7)[45]: "Values of the knowledge culture are trust, collaboration, openness, autonomy, learning receptivity and care" (Sollberger 2006, p. 119)[46]. These dimensions play a central role in the implementation and integration of a holistic knowledge management, including the absorption of external knowledge. Thereby, it depends on the degree of the shape of the several dimensions whether the processes of knowledge management and knowledge absorption are supported, made more difficult or even made impossible (cf. Sollberger 2006, p. 116). Consequently, organizational culture is a critical factor in the absorption of external knowledge.

Knowledge Culture					
Trust	Collabo-ration	Openness	Autonomy	Learning receptivity	Care

Figure 7 Dimensions of a Knowledge Culture[47]

Different studies emphasize different dimensions with regard to the work of *Sollberger* (2006). An exemplary overview of various approaches to the dimensions of organizational culture are as follows[48]: According to *Allee* (1997), trust

---

[45] The construct of organizational culture developed by *Sollberger* (2006), including all of its dimensions, is described in detail in section 3.2.2, which explains the parameters of the model of external knowledge absorption by SMEs.
[46] Translated by the author.
[47] Author's own figure; the terms 'knowledge culture' and 'knowledge-friendly organizational culture' are used synonymously throughout this thesis.
[48] The entire literature review can be found in *Sollberger* (2006, p. 115 ff.) and a summary of the characteristics of knowledge culture with the corresponding literature can be found in the table in *Sollberger* (2006, p. 115 ff.).

and openness are key elements of a knowledge culture in an environment where constant learning is highly appreciated and supported (cf. Allee 1997, p. 212). *Pemberton & Stonehouse* (2002) focus on aspects that are very important for knowledge culture with regard to an organization's position and knowledge management process: "Reliance on trust, openness and individual responsibility, the main themes of the arguments, proposed, is not without its pitfalls, but provided the cultural environment, structure, infrastructure and incentives are in place, the potential for advancement clearly exists, with individual empowerment representing an essential vehicle for enhancing an organisation's position and knowledge management prowess" (Pemberton & Stonehouse 2002, p. 89). *Robbins* (2003) emphasizes learning and the ongoing improvement: "It is a culture in which people are incentivized to share knowledge with their colleagues and leverage the learning of others rather than to hoard information and reinvent wheels. It is a culture that is committed to improving itself and the businesses it serves" (Robbins 2003, online). According to *Ahmed, Kok & Loh* (2002), knowledge management is essentially about the sharing of knowledge; they highlight that trust, openness and collaboration build the basic foundations of sharing (cf. Ahmed et al. 2002, p. 62). Lee & Choi (2003) confirm "that knowledge creation is associated with cultural factors such as collaboration, trust, and learning" (Lee & Choi 2003, p. 210) and give the example by *Huemer, Krogh & Johan* (1998) that "groups are most creative when their members collaborate; members stop holding back when they have mutual trust" (Lee & Choi 2003, p. 210).

Based on the conceptual principles of absorptive capacity and organizational culture in the following section 3 a model of external knowledge absorption is developed.

3 A Model of External Knowledge Absorption

The purpose of the model is to describe the absorption of external knowledge and to explain the influence of organizational culture to derive starting points for organizational-culture-driven control of external knowledge absorption. Furthermore, an empirical test of the model should be possible, so that the theoretically identified starting points for how a knowledge-oriented organizational culture should be designed to support the absorption of external knowledge can be realized in terms of their practical suitability.

Modeling of the model of external knowledge absorption by SMEs is carried out in three steps. In the first step in section 3.1, theories informing absorptive capacity are introduced to obtain a deeper understanding of the concept of absorptive capacity: organizational learning, innovation, dynamic capabilities, knowledge-based view, managerial cognition and coevolution. In the second step, in section 3.2, the parameters of the model that are relevant to the identification of starting points of a cultural design concept for external knowledge absorption, particularly in the organizational culture context, are worked out: absorptive capacity and organizational culture. With regard to absorptive capacity the four capabilities of absorptive capacity, 'acquisition', 'assimilation', 'transformation' and 'exploitation' are explained in detail in section 3.2.1. With respect to organizational culture, the six dimensions of organizational culture — 'trust', 'collaboration', 'openness', 'autonomy', 'learning receptivity' and 'care' — are explained in detail in section 3.2.2. In a third step after explaining the parameters of the model in section 3.3 the relationships between the dimensions of organizational culture and the capabilities of absorptive capacity are worked out and be transferred into an empirically testable model in a third step.

3 A Model of External Knowledge Absorption

3.1 Absorptive Capacity in Organizational Theories

Cohen & Levinthal (1989) put R&D at the center of firms' innovative processes and present learning and innovation as the two faces of R&D (cf. Cohen & Levinthal 1989, p. 569 ff.)[49]. They position absorptive capacity as a key construct in the literature on learning and innovation management and lay the groundwork for theoretical contributions, constructs and implications in subsequent years (cf. Volberda et al. 2010, p. 932 ff.).

In line with past research on absorptive capacity (e.g., cf. Flatten et al. 2011a, cf. Flatten et al. 2011b, cf. Jansen, Van den Bosch & Volberda 2005 cf. Liao et al. 2003), and as previously mentioned, this thesis follows the reconceptualization of the construct of absorptive capacity offered by *Zahra & George* (2002). With their reconceptualization, *Zahra & George* (2002) make three contributions to the literature that support the role of learning and innovation for absorptive capacity but show that absorptive capacity is informed by other theories in addition to learning and innovation (cf. Zahra & George 2002, p. 185 f.):

1. They recognize absorptive capacity as a dynamic capability that influences the competitive advantage of a firm. Therefore, dynamic capabilities are discussed in section 3.1.3.
2. They show that absorptive capacity as a dynamic capability can be changed when the knowledge-based assets of a firm are redefined and deployed through managerial actions. Therefore, the knowledge-based view is discussed in section 3.1.4 and managerial cognition is discussed in section 3.1.5;
3. They broaden the theoretical interpretation of the absorptive capacity construct when they assumed that dynamic capability influences the creation of other organizational competencies and provides firms with strategic flexibility, innovation and performance as sources of competi-

---

[49] Learning is explained in detail in section 3.1.1 and innovation is explained in detail in section 3.1.2. Both are theories that inform absorptive capacity.

tive advantage. This supports the role of learning and innovation and additionally clarifies the role of coevolution, which are described in section 3.1.6.

In summary, absorptive capacity is informed by

- **organizational learning** (e.g., Cohen & Levinthal 1989, Lane, Salk & Lyles 2001, Lane et al. 2001, Reagans & McEvily 2003),
- **innovation** (e.g., Cohen & Levinthal 1989, Feinberg & Gupta 2004, Rothaermel & Alexandre 2009, Benson & Ziedonis 2009),
- **dynamic capabilities** (e.g., Cohen & Levinthal 1990, Zahra & George 2002, Jansen et al. 2005, Lichtenthaler 2009),
- **the knowledge-based view** (e.g., Kogut & Zander 1992, Grant's 1996a, Grant 1996b, Matusik & Heeley 2005),
- **managerial cognition** (e.g., Prahalad & Bettis 1986, Bettis & Prahalad 1995, Minbaeva et al. 2003, Lenox & King 2004) and
- **coevolution** (e.g., Cohen & Levinthal 1994, Lewin & Volberda 1999, Van den Bosch, Volberda & De Boer 1999, Volberda & Lewin 2003) (cf. Figure 8; cf. Volberda et al. 2010).

The following sections present a review of absorptive capacity according to each of the six theories.

3 A Model of External Knowledge Absorption   57

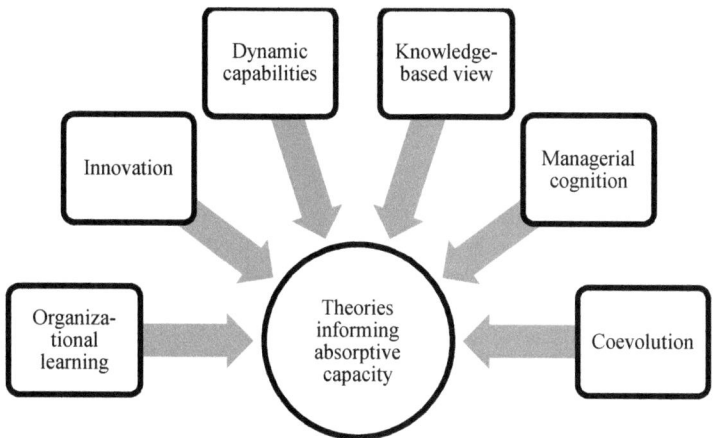

Figure 8 Theories Informing Absorptive Capacity[50]

3.1.1 Organizational Learning

Organizational learning is an area within organizational theory that studies models and theories about the way an organization learns (cf. Kieser & Ebers 2006, p. 185). "Organizational learning theory suggests that the generation of new organizational knowledge is maximized in domains close to the domain of existing knowledge, in conditions under which there are few existing organizational routines to unlearn and organizational assimilation and subsequent retrieval of the knowledge occurs in an intense and repetitive fashion" (Autio, Sapienza & Almeida 2000, p. 911). This suggestion emphasizes an organization's knowledge and organizational learning. Organizational knowledge is defined as "its capacity to apprehend and use relationships among critical factors in such a way to achieve intended ends" (Autio et al. 2000, p. 911) and organizational learning is defined as "the process of assimilating new knowledge into the organization's knowledge base" (Autio et al. 2000, p. 911).

---

[50] Author's own figure.

Therefore, new knowledge is only absorbed when it is assimilated into organizational routines and processes.

Following arguments based on absorptive capacity, *Reagans & McEvily* (2003) state that the higher the similar training and background characteristics of employees, the higher the probability that knowledge is transferred between them (cf. Reagans & McEvily 2003, p. 243). With this statement, they support the argument of *Cohen & Levinthal* (1990) that the potential of a firm to learn is determined by prior related knowledge[51]. The best possible learning and knowledge-assimilating effect is possible in domains close to the existing knowledge base of a firm: *Autio et al.* (2000) highlight that "the more similar the prior knowledge is to new knowledge, the easier the absorption of the new knowledge" (Autio et al. 2000, p. 911).

The roots of absorptive capacity are found in the organizational learning literature of the 1980s, e.g., *Fiol & Lyles* (1985) and *Cohen & Levinthal* (1989). As already mentioned at the beginning of section 3.1, *Cohen & Levinthal* (1989) strongly link absorptive capacity to learning and innovation and performance of firms. They argue that the recognition of the role of R&D as an enhancer of a firm's ability to assimilate and exploit existing knowledge suggests "that the ease and character of learning within an industry will both affect R&D spending and condition the influence of appropriability and technological opportunity conditions on R&D" (Cohen & Levinthal 1989, p. 569). One year later, *Cohen & Levinthal* (1990) argue that absorptive capacity is critical to innovative capabilities and that the ability to evaluate and utilize outside knowledge is largely a function of a firm's level of prior related knowledge (cf. Cohen & Levinthal 1990, p. 129). They also develop the construct of absorptive capacity through an examination of the cognitive structures that underlie learning.

---

[51] The antecedents of absorptive capacity are explained in section 2.1.2.

3 A Model of External Knowledge Absorption 59

Lyles & Salk (1996) and *Lane, Salk & Lyles* (2001) investigate not only absorptive capacity in an interorganizational context but also the absorptive capacity of each involved party. Therefore, their investigations are interesting for this review. *Lyles & Salk* (1996) investigate how organizational characteristics, structural mechanisms and contextual factors influence the acquisition of knowledge from the foreign parent in an IJV (international joint venture). The capacity to absorb knowledge is one of these organizational IJV characteristics (cf. Lyles & Salk 1996, p. 881). Therefore, they investigate the absorptive capacity of the involved ventures and find that the absorptive capacity of IJV organizations has a strong relationship to the ability to assimilate and apply new knowledge (cf. Lyles & Salk 1996, p. 898). Furthermore they find "that capacity to learn, mainly the flexibility, creativity and knowledge about employees, is a significant indicator of knowledge acquisition form the foreign parent [...] [and with regard to ownership and conflict] that differences in the IJV ownership structure affect at least some aspects of knowledge acquisition" (Lyles & Salk 1996, p. 896). Furthermore, with respect to knowledge acquisition and performance, their results suggest knowledge acquisition and that each of the performance measures positively influence one another (cf. Lyles & Salk 1996, p. 897).

In summary, it is important for a deeper understanding of absorptive capacity that a firm's learning potential is determined by prior related knowledge because the higher the employees' prior related knowledge, the higher the probability that they will acquire knowledge. The best possible learning and knowledge assimilating effect occurs when there is a high level of prior related knowledge. With respect to transformation and exploitation from an organizational learning point of view, it is important that the generation of new knowledge be maximized in domains close to the domain of prior related knowledge. Furthermore, it is very important for transformation and exploitation that the subsequent retrieval of the knowledge occur in an intense and repetitive way. Therefore, corresponding routines and processes are necessary: For transformation it is necessary to develop and refine those routines that facilitate combining existing knowledge with acquired and assimilated knowledge for future use. For exploitation, it is necessary to refine, extend, and leverage existing routines, competencies and technologies or to create new ones by incorporating acquired and transformed knowledge into a firm's own operations.

3.1.2 Innovation

Many investigations of absorptive capacity build on insights from the innovation literature. Starting with the findings of *Cohen & Levinthal* (1989), firms not only invest in R&D to pursue new processes and product innovation but also to develop and maintain their absorptive capacity to assimilate, transform and exploit external knowledge (cf. Cohen & Levinthal 1989, p. 593). Therefore, firms target not only the generation of innovation but also the maintenance of their absorptive capacity to generate long-term competitive advantages. *Cohen & Levinthal* (1989) link absorptive capacity strongly to both process and product innovation and learning.

Companies must respond to the process and product innovation imperative and are increasingly moving from a traditional, closed model of innovation (closed innovation) to an open model of innovation (open innovation). In the closed innovation model, companies generate, develop and commercialize their own knowledge and ideas, and in the open innovation model, knowledge and ideas flow into and out of companies because organizational boundaries to innovation are porous (cf. Morris, Kuratko & Covin 2008, p. 97).

The term 'open innovation' was coined by Henry William Chesbrough to describe a paradigm shift: He calls the old paradigm 'closed innovation', in which companies are strongly self-reliant in terms of creating knowledge and introducing it to the market in the form of new products and services because they cannot be sure of the quality, availability and capability of others' ideas (Chesbrough 2006, p. xx). 'Open innovation' can be seen as "the use of purposive inflows and outflows of knowledge to accelerate internal innovation, and expand the markets for external use of innovation, respectively" (Chesbrough,

Vanhaverbeke & West 2008, p. 1). It has been defined as "a paradigm that assumes that firms can and should use external ideas as well as internal ideas, and internal and external paths to market, as the firms look to advance their technology" (Chesbrough 2006, p. xxiv). This model is termed 'open' innovation because there are many ways for ideas to flow into the process, and many ways for them to flow out to the market. The possibilities for utilizing knowledge on the market include licensing the knowledge for another firm's market, technology spin-offs for new markets and going to market using a company's own marketing and sales channels (cf. Chesbrough et al. 2008, p. 3). Table 7 lists closed innovation principles and contrasts them to open innovation principles.

Table 7 Closed Innovation Principles Versus Open Innovation Principles[52]

Closed Innovation Principles	Open Innovation Principles
The smart people in our field work for us.	Not all the smart people work for us. We need to work with smart people inside and outside our company.
To profit from R&D, we must discover it, develop it and ship it ourselves.	External R&D can create significant value; internal R&D is needed to claim some portion of that value.
If we discover it, we will get it to market first.	We do not have to originate the research to profit.
The company that gets an innovation to market first will win.	Building a better business model is better than getting to market first.
If we create the most and the best ideas in the industry, we will win.	If we make the best use of internal and external ideas, we will win.
We should control our IP, so that our competitors do not profit from our ideas.	We should profit from others' use of our IP, and we should buy others' IP whenever it advances our own business model.

---

[52] Author's own table, referencing *Chesbrough* (2006, p. xxvi).

Rigby & Zook (2002) offer four reasons why companies are increasingly choosing to pursue an approach that follows open innovation principles (Rigby & Zook 2002, p. 82 ff.):

1. "Importing new ideas is a good way to multiply the building blocks of innovation" (Rigby & Zook 2002, p. 82), meaning companies can potentially offer more and better outputs by accessing external inputs. This is because if people have more ideas to choose from and different knowledge, the cost, quality and speed of innovation improve.
2. "Exporting ideas is a good way to raise cash" (Rigby & Zook 2002, p. 83), meaning a company's ideas can have market value that is exploitable through its sale, licensing, etc., to other companies.
3. "Exporting ideas gives companies a way to measure an innovation's real value and to ascertain whether further investment is warranted" (Rigby & Zook 2002, p. 84), meaning offers to sell internally developed ideas to external markets can be a test of the true market value of those ideas.
4. "Exporting and importing ideas helps companies clarify what they do best" (Rigby & Zook 2002, p. 84) because companies often think that their core business is broader than it is in reality and market knowledge helps them to discover where they are stronger and weaker than assumed.

The open innovation model is an approach to address the challenges of innovation, but "It's a supplement to the steps that are needed to resolve the basic problem of innovation, not a solution in itself. The fundamental problem in innovation isn't one of finding more new ideas: it's a matter of establishing a way of running the organization that is open to exploring new ideas and willing to back the most promising of them with resources and talent" (Denning 2005, p. 8). This problem is very important with regard to the transformation and exploitation capabilities. Transformation and exploitation build on the refining and leveraging of routines and therefore, there are ways of running an organization that can be problematic for innovation, as has been pointed out by *Denning* (2005).

In summary, it is important for absorptive capacity that from an innovation point of view, firms can and should use both external and internal knowledge to accelerate internal innovation and to expand their markets for the external use of innovation. Companies should follow an open innovation model. They should acquire and assimilate external knowledge, and external knowledge and ideas should flow into companies because they can potentially offer more and better outputs by accessing external inputs. Open innovation combines internal and external ideas and gives strong support to the market introduction of internal ideas through external channels to generate additional value and competitive advantage. The key to managing knowledge for open innovation is to enhance concepts capable of supporting new forms of collaboration, knowledge management and absorption. With regard to the full absorption of knowledge, it is important to note that it is not missing new ideas but rather establishing a way of running a company that is open to exploring new ideas that is the problem. Once again — as with organizational learning to ensure the subsequent retrieval of knowledge in an intense and repetitive way — corresponding routines and processes are necessary.

3.1.3 Dynamic Capabilities

Dynamic capabilities are defined "as the firm's ability to integrate, build, and reconfigure internal and external competences to address rapidly changing environments" (Teece et al. 1997, p. 516). Therefore, they can be seen as an integrative approach to understand newer sources of competitive advantage (cf. Teece et al. 1997, p. 510).

At the beginning of the 1990s, *Cohen & Levinthal* (1990) had already highlighted that an organization's absorptive capacity depends on links across individual capabilities (cf. Cohen & Levinthal 1990, p. 133)[53]. Later, they argue that sustaining this absorptive capacity over time requires investments in the dynamic capability but results in an ability to permit firms to exploit valuable opportunities and better envision their emergence (cf. Cohen & Levinthal 1994, p. 244).

When *Zahra & George* (2002) define absorptive capacity as "the set of organizational routines and processes by which firms acquire, assimilate, transform, and exploit knowledge to produce a dynamic capability" (Zahra & George 2002, p. 186), they extend the dynamic nature and introduce a dynamic capabilities perspective of absorptive capacity. They distinguished among four capabilities of absorptive capacity that constitute potential and realized absorptive capacity. *Jansen et al.* (2005) build on *Zahra & George* (2002) and investigate how organizational antecedents affect the managing of the potential and realized absorptive capacity. They note that "organizational mechanisms associated with coordination capabilities (cross-functional interfaces, participation in decision making, and job rotation) primarily enhance a unit's potential absorptive capacity [...] [and that] organizational mechanisms associated with socialization capabilities (connectedness and socialization tactics) primarily increase a unit's realized absorptive capacity" (Jansen et al. 2005, p. 999). This understanding allowed them to identify reasons that units find it difficult to manage potential and realized absorptive capacity and why they vary in their ability to create value from their absorptive capacity (cf. Jansen et al. 2005, p. 999).

Lichtenthaler (2009) identifies technological and market knowledge as two critical components of prior knowledge in the organizational learning processes of absorptive capacity and discovered that exploratory, transformative, and exploitative learning have complementary effects on innovation and per-

---

[53] *Cohen & Levinthal* (1990, p. 133), referencing *Nelson & Winter* (1982).

3 A Model of External Knowledge Absorption 65

formance under different environmental conditions (cf. Lichtenthaler 2009, p. 822). Therefore, he draws inferences from the multidimensional nature of absorptive capacity about firm discrepancies in profiting from external knowledge and works out the importance of dynamic capabilities in contexts characterized by high degrees of technological and market turbulence (cf. Lichtenthaler 2009, p. 822).

In summary, from a dynamic capabilities point of view, dynamic capabilities are inevitable for the absorptive capacity because they constitute a firm's ability to integrate, build and reconfigure competences. These competences enable firms to develop routines and processes that allow them to analyze, process, interpret, and understand external sources of acquired knowledge, to develop and refine those routines that facilitate combining existing knowledge with acquired and assimilated knowledge for future use and to refine, extend, and leverage existing routines, competencies and technologies or to create new ones by incorporating acquired and transformed knowledge into their own operations. The first competence means to assimilate knowledge, the second means to transform knowledge and the third means to exploit knowledge.

Sustaining absorptive capacity over time requires investments in dynamic capability, but those investments result in an ability that permits firms to exploit valuable opportunities and stay competitive.

3.1.4 Knowledge-Based View

Kogut & Zander (1992) have considered knowledge as a firm's most important resource, because it is the main determinant of competitive advantage: "In our view, the central competitive dimension of what firms know how to do is to create and transfer knowledge efficiently within an organizational context" (Kogut & Zander 1992, p. 384). This view influences the relevance of absorptive capacity because absorptive capacity is the key to developing and increasing a firm's knowledge base (cf. Volberda et al. 2010, p. 935).

In *Grant's* (1996a) knowledge-based theory of the firm, knowledge is even referred to as "the most strategically important of the firm's resources" (Grant 1996a, p. 110). *Grant* (1996b) has developed a knowledge-based theory of organizational capability and has analyzed mechanisms through which knowledge is integrated a firm to create a capability of exploring its potential to establish competitive advantage in dynamic market settings, where market leadership and power are continually undermined by competition and external change (cf. Grant 1996b, p. 375 ff.). He has shown that the processes and routines through which a company absorbs specialized knowledge are fundamental to its ability to create and sustain competitive advantage (cf. Grant 1996b, p. 384). *Van den Bosch et al.* (1999) share that view and have stated that firms need increased absorptive capacity due to a more demanding knowledge environment (cf. *Van den Bosch et al.* 1999, p. 551 ff.).

Matusik & Heeley (2005) have investigated how multiple dimensions (a firm's relationship to its external environment, the structure, routines, and knowledge base of the main value-creating group(s) and individuals' absorptive abilities) embedded in the absorptive capacity construct play a role in influencing firm knowledge and knowledge creation activities because it is very important to understand how firms can effectively absorb and assimilate external knowledge in the context of increased knowledge-based competition (cf. Matusik & Heeley 2005, p. 549 ff.). Those authors have shown that each of the multiple dimensions contributes to increased knowledge or knowledge-creation activities.

In summary, it is very important to absorptive capacity that from a knowledge-based view, the main determinant of competitive advantage is a firm's ability to absorb knowledge and to develop and increase its knowledge base. Knowledge is the most strategically important of a firm's resources, and firms have a great desire for external knowledge that has the potential to complement their existing knowledge bases. Although the knowledge-based view may be considered a general tenet of recent management theory, the notion of absorptive capacity specifically draws upon innovation-related outcomes of the knowledge-based view (cf. Roberts, Galluch, Dinger &Grover 2012). The processes and routines through which a company assimilates, transforms and exploits knowledge are fundamental to its ability to create innovation and sustain competitive advantage.

3.1.5 Managerial Cognition

Managers are assumed to be 'information workers' (cf. McCall & Kaplan 1985) and "a manager is a craftsperson whose material is information" (McCall & Kaplan 1985, p. 14). Following the data–information–knowledge–wisdom hierarchy set forth by *Rowley* (2007), "data can be used to create information; information can be used to create knowledge, and knowledge can be used to create wisdom" (Rowley 2007, p. 164). Therefore, managers build the basis for knowledge.

Walsh (1995) has described information workers' tasks as follows: Information workers "spend their time absorbing, processing, and disseminating information about issues, opportunities, and problems" (Walsh 1995, p. 280)[54]. Therefore, they perceive information through their own cognitive lenses (cf. Volberda et al. 2010, p. 933). This means that managers can reduce complexity by using mental maps developed through experience in their core businesses, which sometimes are inappropriately applied to other businesses (cf. Prahalad & Bettis 1986, p. 485). So-called 'dominant general management logic' (or dominant logic) consists of these mental maps (cf. Prahalad & Bettis 1986, p. 485 ff.). The dominant logic can have an impact on new organization forms and also have consequences for absorptive capacity because firms that apply classical

---

[54] *McCall & Kaplan* (1985) have explained this thought in more detail: "Most managers spend most of their time with other people, and with these people they go for about the business of exchanging information (McCall et al., 1978). What little time managers get to themselves, uninterrupted by people or calls, they spend reading (absorbing information), thinking (processing information), and writing (disseminating information)" (McCall & Kaplan 1985, p. 14).

management logic are likely to favor a functional form and system capabilities, which in turn limit their capacity to absorb knowledge (cf. Van den Bosch et al. 1999, p. 560).

With regard to organizational forms, *Lyles & Schwenk* (1992) have noted the increased interest shown by strategic management research in the cognition of top management teams and go on to discuss the existence and maintenance of organizational knowledge structures (cf. Lyles & Schwenk 1992, p. 170). They have discovered that organizational knowledge structures of organizational forms are built out of a social process (cf. Lyles & Schwenk 1992, p. 170 f.). The provision of information within a knowledge structure is considered as an important factor influencing managerial cognition and absorptive capacity (cf. Lenox & King 2004, p. 332 ff.). Managers themselves may lack of the necessary knowledge to efficiently discover and absorb new practices as a critical element of sustained competitive advantage (cf. Lenox & King 2004, p. 332). Thus, more knowledge acquired from external sources may lead to a higher degree of managerial awareness and cognition. Furthermore, managers can directly, positively affect the absorptive capacity of their firms when they provide knowledge to potential adopters within those firms (cf. Lenox & King 2004, p. 332). With regard to this knowledge provision, *Lenox & King* (2004) have explored the extent to which managers can support the development of absorptive capacity by directly providing information to agents in the firm that might adopt new practices. They find "that the effectiveness of managerial information provision depends on the degree to which potential adopters have information from other sources[…] [and] that information from previous adopters and past events reduces the effect of information provision, while experience with related practices amplifies it" (Lenox & King 2004, p. 331). Therefore, the effectiveness of managerial information provision is contingent on the recipient of the knowledge.

Following the idea that the organizational form, increasingly characterized by flexibility and adaptability, is a management tool in the alignment of organization and environment, several authors have previously investigated

3 A Model of External Knowledge Absorption

managerial cognition and absorptive capacity. For example, *Bettis & Prahalad* (1995) have elaborated how the concept of *Prahalad & Bettis* (1986) has been further developed in recent years and have focused on the following points: dominant logic as an information filter which is then incorporated into an organization's strategy, systems, values, expectations, and reinforced behavior (cf. Bettis & Prahalad 1995, p. 7); dominant logic as a level of strategic analysis (cf. Bettis & Prahalad 1995, p. 8 f.); the unlearning (forgetting) curve (cf. Bettis & Prahalad 1995, p. 9 f.); dominant logic as an emergent property of organizations as complex adaptive systems (cf. Bettis & Prahalad 1995, p. 14); and the relationship between organizational stability and dominant logic (cf. Bettis & Prahalad 1995, p. 12 ff.). *Bettis & Prahalad* (1995) have come to the following conclusion: "The dominant logic seems to fit comfortably into the domain of emergent properties of complex adaptive systems" (Bettis & Prahalad 1995, p. 14). Furthermore, *Bettis & Prahalad* (1995) have discussed the questions of why many institutions find it so difficult to change and why many institutions see change in their environments but are unable to act. They have summed up as follows: "Often the focus in trying to answer such questions has been on the surface architecture of the organization strategy, structure, and systems instead of underlying structures and foundations, such as the dominant logic, that support the visible features. We believe that the concept of dominant logic can be useful in developing a much more thorough understanding of these issues" (Bettis & Prahalad 1995, p. 7).

In summary, it is very important for absorptive capacity that from a managerial recognition point of view, managers can reduce the complexity of absorbing new knowledge. This facilitates the development and refinement of routines and processes, e.g., of those routines that facilitate combining existing knowledge with acquired and assimilated knowledge for future use. The dominant logic can have consequences for absorptive capacity because firms that apply classical management logic will limit their capacity to absorb knowledge.

3.1.6 Coevolution

The framework of *Cohen & Levinthal* (1994) has made the following suggestion: "Managers are mistaken if they believe that in the face of emerging technologies, uncertainty is best resolved by waiting passively for clearer signals from the environment. Even if a firm correctly believes itself to be the only one capable of exploiting a new technology, not investing in the requisite absorptive capacity will result in a substantial opportunity cost to itself and society" (Cohen & Levinthal 1994, p. 245; own citation). Therefore, firms can benefit from investigations of their absorptive capacity to preempt environmental changes. Absorptive capacity is a critical factor in industrial competitiveness and enables firms not only to exploit external knowledge but also to predict more accurately the nature of future technological advances (cf. Cohen & Levinthal 1994, p. 227).

Lewin, Long & Carroll (1999) have hypothesized "that firm strategic and organization adaptations coevolve with changes in the environment [...] and organization population and forms, and that new organizational forms can mutate and emerge from the existing population of organizations" (Lewin et al. 1999, p. 535). Their alternative theory of organization-environment coevolution "considers organizations, their populations, and their environments as the interdependent outcome of managerial actions, institutional influences, and extra-institutional changes [...] [and] incorporates potential differences and equifinal outcomes related to country-specific variation" (Lewin et al. 1999, p. 535). Absorptive capacity influences relationships among managerial actions, institutional influences and extra-institutional changes and organizations, their populations and their environments. *Van den Bosch et al.* (1999) have investigated how organizational forms and combinative capabilities, as two specific organizational dimensions of absorptive capacity, influence levels of absorptive capacity and therefore, levels of prior related knowledge (cf. van den Bosch et al. 1999, p. 551). To explain how knowledge environments coevolve with the emergence of organization forms and combinative capabilities that are suitable for absorbing knowledge, those authors have developed a framework in which absorptive capacity is related to both micro- and macro-coevolutionary effects (cf. van den Bosch et al. 1999, p. 551). Based on *Cohen & Levinthal* (1994), they have suggested that firms with a higher level of absorptive capacity will tend to be more proactive and are therefore better at anticipating the emergence of valuable developments because proactive firm behavior is very important in turbulent environments (cf. van den Bosch et al. 1999, p. 552). *Huygens, Baden-Fuller, van den Bosch & Volberda* (2001) have identified a research gap in the understanding of co-evolutionary processes within the field of strategic management and have investigated the co-evolution of capabilities and competition within the competitive environment with case studies: "The case studies show how these firms increased their absorptive capacity, not only by increasing the level of prior related knowledge, but also by deliberately changing their organization form and combinative capabilities" (Huygens et al. 2001, p. 552). *Volberda & Lewin* (2003) have identified four co-evolutionary generative mechanisms (naive selection, managed selection, hierarchical renewal and holistic renewal) that clarify the extensive range of evolutionary paths that can take place in the workforce (cf. Volberda & Lewin 2003, p. 2111). Furthermore, they have noted that "the managed selection engine provides the foundations of the underlying principles of co-evolving self-renewing organizations: managing internal rates of change, optimizing self-organization, and balancing concurrent exploration and exploitation" (Volberda & Lewin 2003, p. 2111).

In summary, from a coevolution point of view, it is very important for absorptive capacity that firm strategic and organization adaptations — e.g., of the processes and routines during the assimilation, transformation and exploitation of knowledge — coevolve with changes in the environment because therefore, absorptive capacity coevolves with changes in the environment. This absorptive capacity is critical for competitive advantage because it enables firms to predict more accurately the nature of future technological advances: It is very important for companies not to wait passively for clearer signals from the environment, but rather, they must actively acquire necessary knowledge corresponding to the needs of recognized future technological advances. Firms with a higher level of absorptive capacity are more proactive and therefore are better at anticipating the emergence of valuable environmental developments because currently, proactive firm behavior is very important in turbulent environments, where many SMEs are confronted with rough market conditions.

3.1.7 Summary of Theories Informing Absorptive Capacity

There is a high level of heterogeneity in the research on absorptive capacity, which is an indication of the richness of that construct (cf. Volberda et al. 2010, p. 936). The literature review of absorptive capacity in organizational theory shows that the construct of absorptive capacity has achieved acclaim in various organizational theories. Although many authors' investigations are still embodied in the themes of learning and innovation, absorptive capacity research has also addressed managerial cognition, the knowledge-based view, dynamic capabilities and coevolution. The results of the literature review indicate that each of the theories informing absorptive capacity plays an important role in determining whether a firm is able to recognize the value of external knowledge and to assimilate, transform and exploit it — to absorb knowledge — or not.

Because increased organizational learning enlarges a firm's knowledge base and therefore its absorptive capacity, firms should develop their learning capacities to enable their acquisition of external knowledge. Both organizational learning and absorptive capacity are crucial for the successful implementation of open innovation principles because the success of open innovation is determined by absorptive capacity. The higher the absorptive capacity, the higher the ability to recognize the value of external knowledge, to assimilate it and to apply it to commercial ends.

Managerial cognition can influence absorptive capacity. Managers can reduce complexity by using mental maps, but on the other side, the comprehension of external knowledge through mental maps can cause a firm to lose valuable knowledge, which may lower the knowledge base and absorptive capacity. Following the knowledge-based view, because firms greatly desire external

3 A Model of External Knowledge Absorption

knowledge that has the potential to complement their existing knowledge bases, this loss must be avoided because knowledge is most strategically important of a firm's resources. Furthermore, firms must pay attention to rapidly changing environments, because they invest in the dynamic capability necessary for absorptive capacity and therefore, concern coevolution.

From the perspective of several theories, the hypotheses of section 3.3 are theoretically supported, but before these hypotheses are discussed, section 3.2 addresses the parameters of the model.

3.2 The Parameters of the Model

The sections that follow describe several parameters of the model of external knowledge absorption. First, section 3.2.1 explains the construct of absorptive capacity and its four capabilities: acquisition, assimilation, transformation and exploitation. Second, section 3.2.2 introduces the dimensions of the organizational culture, following *Sollberger* (2006): trust, collaboration, openness, autonomy, learning receptivity and care.

3.2.1 The Construct of Absorptive Capacity

The following sections provide a detailed explanation of the four capabilities of absorptive capacity — 'acquisition', 'assimilation', 'transformation' and 'exploitation' — beginning with the acquisition capability.

3.2.1.1 The Acquisition Capability

Zahra & George (2002) have defined acquisition as "a firm's capability to identify and acquire externally generated knowledge that is critical to its operations"

(Zahra & George 2002, p. 189). *Flatten et al.* (2011a) have noted the external character of knowledge sources by defining acquisition as "a firm's ability to identify and obtain knowledge from external sources" (Flatten et al. 2011a, p. 100). In summary, acquisition refers to a firm's ability to identify and acquire knowledge relevant to its operations from external knowledge sources (cf. Flatten et al. 2011a, p. 100, cf. Flatten et al. 2011b, p. 138, cf. Zahra & George 2002, p. 189). Therefore, on the one hand knowledge acquisition involves the development of skills, insights and relationships (cf. DiBella & Nevis 1998, p. 87 ff.), and on the other hand, firms need prior related knowledge if they are to acquire new knowledge (cf. Cohen & Levinthal 1990, p. 129).

During the acquisition of external knowledge, the following point is important: "Before an individual can earn an entrepreneurial profit from an opportunity, he or she must discover that it has value" (Shane 2000, p. 451)[55]. Furthermore, *Cohen & Levinthal* (1990) have highlighted that research on memory development shows that a knowledge base increases the ability to put new knowledge into memory, the means of acquiring knowledge, and the ability to recall and use that knowledge (cf. Cohen & Levinthal 1990, p. 129). They have suggested that memory development is self-reinforcing when more objects, patterns and concepts are stored there, which makes it easier for an individual to acquire new information and to use it in new settings (cf. Cohen & Levinthal 1990, p. 129).

With regard to knowledge acquisition routines, the effort expended in such routines has three attributes — intensity, speed and direction — that can influence absorptive capacity (cf. Zahra & George 2002, p. 189). *Zahra & George* (2002) have claimed that the intensity and speed of a firm's efforts to identify and gather knowledge can determine the quality of that firm's acquisition capability (cf. Zahra & George 2002, p. 189). Therefore, they have explained that according to *Clark & Fujimoto* (1991), the ability of a firm to

---

[55] According to *Shane* (2000), search and recognition explain this discovery process (cf. Shane 2000, p. 451; further information on search and recognition can be found in *Shane* (2000).

3 A Model of External Knowledge Absorption 75

achieve speed is limited by difficulties and efforts related to short learning cycles and to assembling the necessary resources to build absorptive capacity (cf. Clark & Fujimoto 1991, p. 206 f.). All activities have a different richness and complexity and thus, firms need competencies across various fields to successfully acquire external knowledge (cf. Rocha 1999, p. 255).

Acquisition Capability

- Identifying new knowledge
- Learning from partners
- Acquiring knowledge relevant to a company's operations
- Prior knowledge as a prerequisite

Figure 9 Detailed Aspects of the Acquisition Capability[56]

Following Figure 9's summary of the detailed aspects of the acquisition capability, section 3.2.1.2 explains the assimilation capability as the second capability of absorptive capacity.

3.2.1.2   The Assimilation Capability

Although *Cohen & Levinthal* (1990) have assumed that externally available knowledge cannot be passively assimilated, *Zahra & George* (2002) have emphasized that "assimilation refers to routines and processes that allow it to analyze, process, interpret, and understand the information obtained from external sources" (Zahra & George 2002, p. 189). To assimilate and use external

---

[56] Author's own figure.

knowledge, SMEs need prior related knowledge (e.g., basic skills, a shared language and knowledge of scientific or technological developments in a given field; cf. Cohen & Levinthal 1990, p. 128 f.). The reason is that "the ability to evaluate and utilize outside knowledge is largely a function of the level of prior related knowledge" (Cohen & Levinthal 1990, p. 129). *Zahra & George* (2002) have highlighted that within an assimilation process, comprehension of externally acquired knowledge can be problematic (e.g., because of different heuristics); they also take up a point that has been made by *Teece* (1981), and have emphasized that "comprehension is especially difficult when the value of knowledge depends on the existence of complementary assets that may not be available to the recipient firm [...], however, [it] promotes knowledge assimilation that allows firms to process and internalize externally generated knowledge" (Zahra & George 2002, p. 190).

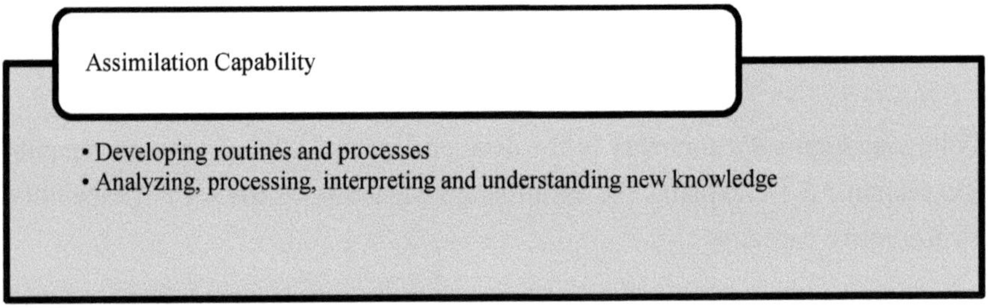

Assimilation Capability

- Developing routines and processes
- Analyzing, processing, interpreting and understanding new knowledge

Figure 10 Detailed Aspects of the Assimilation Capability [57]

Not all assimilated knowledge is used directly. Therefore, *Lichtenthaler* (2009) have highlighted that it is important to practice active knowledge management to keep assimilated knowledge alive, otherwise routines, skills and processes will be lost. For subsequent exploitation, maintained knowledge from the assim-

---

[57] Author's own figure.

3 A Model of External Knowledge Absorption 77

ilation must be reactivated, which can be accomplished by internalization through experience (cf. Argote et al. 2003b, p. 575, cf. Lichtenthaler 2009, p. 825).

Having summarized the detailed aspects of the assimilation capability in Figure 10, section 3.2.1.3 explains the transformation capability, which is the third capability of absorptive capacity.

3.2.1.3 The Transformation Capability
Zahra & George (2002) have reviewed, reconceptualized and extended the original construct of absorptive capacity as set forth by *Cohen & Levinthal* (1990) when they proposed transformation as a further capability of absorptive capacity. In their model, transformation follows assimilation. Transformation refers to a firm's capability to develop and refine those routines that facilitate combining existing knowledge with acquired and assimilated knowledge for future use (cf. Flatten et al. 2011a, p. 100, cf. Zahra & George 2002, p. 190). Although *Cohen & Levinthal* (1990) did not define transformation as a capability of absorptive capacity *Zahra & George* (2002) have broadened the construct of absorptive capacity from the original three capabilities (identify, assimilate, and exploit) by *Cohen & Levinthal* (1990) to four capabilities (acquire, assimilate, transform and exploit).

Spender (1996) has summarized that firms can be viewed as governance structures that enables them to merge, synthesize and transform acquired knowledge into higher-value products and services (cf. Spender 1996, p. 46). During the transformation, existing and new knowledge are combined (cf. Flatten et al. 2011b, p. 138 f.). A firm's transformation capability is represented by "the ability of firms to recognize two apparently incongruous sets of information and then combine them to arrive at a new schema" (Zahra & George 2002, p. 190). Therefore, transformation is accomplished when knowledge is added, deleted, or simply interpreted differently, thereby changing the character of that knowledge. This change is called 'bisociation'.

The underlying pattern of bisociation is "the perceiving of a situation or idea [...] in two self-consistent but habitually incompatible frames of reference" (Koestler 1964, p. 35). Transformation capabilities "through the process of bisociation help firms to develop new perceptual schema or changes to existing processes" (Zahra & George 2002 195). Therefore, it is possible to understand situations and ideas that initially were perceived as incompatible with current cognitive frames of reference (cf. Todorova & Durisin 2007, p. 778). Furthermore, this transformation capability shapes the entrepreneurial mindset (cf. McGrath & McMillan 2000) and fosters entrepreneurial action (cf. Smith & Di Gregorio 2006). As a result of this change, the gain of new insights by the members of an organization enhances its competencies and facilitates its recognition of opportunities (cf. Zahra & George 2002, p. 190).

Transformation Capability

- Developing and refining routines and processes
- Combining existing knowledge with acquired and assimilated knowledge for future use

Figure 11 Detailed Aspects of the Transformation Capability [58]

After having summarized detailed aspects of the transformation capability in Figure 11, section 3.2.1.4 explains the exploitation capability as the fourth capability of absorptive capacity.

---

[58] Author's own figure.

3.2.1.4 The Exploitation Capability

Exploitation refers to a firm's capability to refine, extend and leverage existing routines, competencies and technologies or to create new ones by incorporating acquired and transformed knowledge into their own operations (cf. Flatten et al. 2011a, p. 100, cf. Zahra & George 2002, p. 190). *Cohen & Levinthal* (1990) have already emphasized the application of knowledge in their definition of absorptive capacity, namely "the ability of a firm to recognize the value of new, external information, assimilate it, and apply it to commercial ends" (Cohen & Levinthal 1990, p. 128). According to *Minbaeva et al.* (2003), "the key element in knowledge transfer is not the underlying (original) knowledge, but rather the extent to which the receiver acquires potentially useful knowledge and utilizes this knowledge in own operations" (Minbaeva et al. 2003, p. 387). Therefore, in exploitation the primary emphasis is on the routines, competencies and technologies that allow firms to exploit knowledge, because the presence of such routines, competencies and technologies allow firms to specifically exploit knowledge over extended periods of time by providing structural, systemic and procedural mechanisms (cf. Flatten et al. 2011b, p. 139, cf. Zahra & George 2002, p. 190).

Zahra & George (2002) have provided the following summary: "The outcomes of systematic exploitation routines are the persistent creation of new goods, systems, processes, knowledge, or new organizational forms" (Zahra & George 2002, p. 190). Exploitation denotes a firm's capacity to create something new by harvesting and incorporating transformed knowledge into its operations. Figure 12 summarizes detailed aspects of the exploitation capability.

Exploitation Capability
• Refining, extending, and leveraging existing routines, competencies and technologies • Creating new routines, competencies and technologies by incorporating transformed knowledge into a firm's own operations • Reflecting

Figure 12 Detailed Aspects of the Exploitation Capability [59]

Having explained the construct of absorptive capacity with its four capabilities, section 3.2.2 explains the dimensions of organizational culture to describe all of the several parameters of the model of external knowledge absorption.

3.2.2 The Construct of Organizational Culture

After a short introduction to the six dimensions of organizational culture of a firm — 'trust', 'collaboration', 'openness', 'autonomy', 'learning receptivity' and 'care' — in section 2.2.3, the following sections explain the six dimensions in more detail, starting in the next section 3.2.2.1 with trust.

3.2.2.1 Trust

Although to date there is no universally accepted definition of trust, several authors have agreed that trust is important in a number of ways, as shown in Figure 13 (cf. Rousseau, Sitkin, Burt & Camerer 1998, p. 394).

---

[59] Author's own figure.

3 A Model of External Knowledge Absorption 81

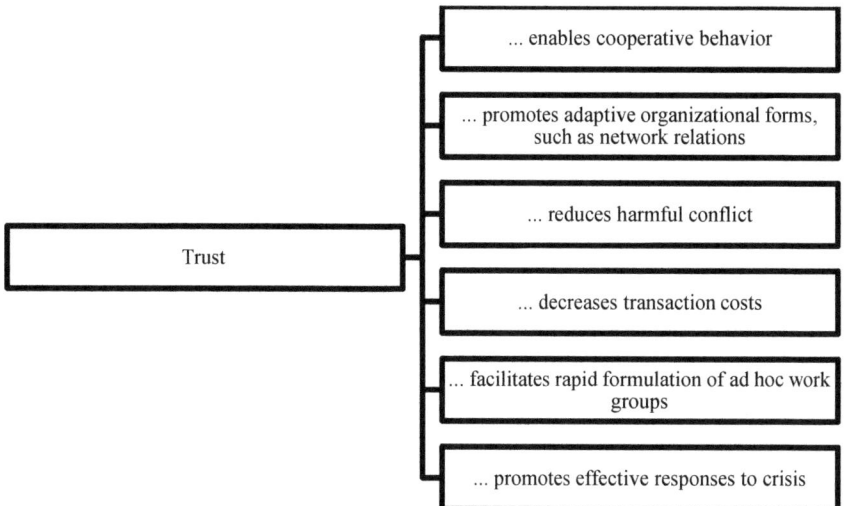

Figure 13 Importance of Trust[60]

Considering this importance, *McEvily Perrone & Zaheer* (2003) have defined trust at a general level as "the willingness to accept vulnerability based on positive expectations about another's intentions or behaviors" (McEvily et al. 2003, p. 92)[61].

In the past, researchers have paid considerable attention to clarifying both the dependence of trust on social contexts and the conditions and dimensions of trust (cf. Jones & George 1998, p. 531). *Jones & George* (1998) have discovered that researchers commonly have viewed trust as "an expression of confidence between the parties in an exchange of some kind-confidence that they will not be harmed or put at risk by the actions of the other party [...] or

---

[60] Author's own figure, referencing *Rousseau et al.* (1998, p. 394). *Rousseau et al.* (1998, p. 394), referencing *Gambetta* (1988), *Meyerson, Weick & Kramer* (1996) and *Miles & Snow* (1992).

[61] *McEvily et al.* (2003, p. 92), referencing *Mayer, Davis & Schoorman* (1995) and *Rousseau et al.* (1998).

confidence that no party to the exchange will exploit the other's vulnerability" (Jones & George 1998, p. 531). Therefore, trust can be a way to compensate for a lack of knowledge about a person, because a person's motives, interests, etc. usually are not the entirety of that person, and a relationship can still be built only with the help of trust (cf. Sollberger 2006, p. 120).

To better understand the processes of selective filtration that occurs in knowledge transfer, *O'Reilly & Roberts* (1974) have investigated the effects of trust under conditions in which a sender has high trust in a receiver and under conditions in which a sender perceives a receiver to have a high level of influence over his future. The investigation has shown the following results: "Unless the sender perceives the receiver as trustworthy, he is unlikely to pass information unfavorable to himself, and, in the downward condition, regardless of trust he does not appear to pass unfavorable-important information" (O'Reilly & Roberts 1974, p. 262). This supports the statement of *McEvily et al.* (2003), namely, "trust has been conceptualized as an expectation, which is perceptual or attitudinal, as a willingness to be vulnerable, which reflects volition or intentionality, and as a risk-taking act, which is a behavioral manifestation" (McEvily et al. 2003, p. 93).

Pemberton & Stonehouse (2002) have demanded a culture that facilitates trust: they have stated that personal responsibility for knowledge dissemination must be encouraged by fostering the degree of trust throughout an organization (cf. Pemberton & Stonehouse 2002, p. 85). Knowledge dissemination is inevitable for the absorption of knowledge. To create a cultural environment that facilitates trust, it is important to note employees' underlying concerns related to sharing knowledge. Companies must avoid a situation in which employees are reluctant to share knowledge or take action because they are afraid that their value as employees and therefore, their job security, are inextricably tied to their personal knowledge and expertise (cf. Davenport et al. 1997, p. 15).

3.2.2.2 Collaboration

Recently, the innovation process has become more complex and knowledge-intensive and networks have enabled a much more rapid flow of knowledge across all types of organizations (cf. Newell, Robertson, Scarbrough & Swan 2009, p. 206). At the same time, innovation itself has become much more open-ended: Linux — as an example of the growth of the 'open-source software movement' — highlights this conflation of design and use and the power of the open innovation approach. From a corporate point of view the cost of controlling a firm's own research and development and its necessary internal competencies has become excessively high (cf. Katzy & Klein 2008, p. 3). "Moreover, time-based competition on a global scale requires extended external collaboration in target markets" (Katzy & Klein 2008, p. 3). Therefore, collaboration with both users and other organizations has become increasingly relevant to innovation for two reasons.

3.2.2.3 Openness

According to *Wathne et al.* (1999), openness can be understood as the "willingness to share knowledge and partner interaction" (Wathne et al. 1999, p. 60). *Stata* (1994) has enhanced the definition of openness and has defined openness as "a willingness to put all the cards on the table, eliminate hidden agendas, make our motives feelings, and biases known, and invite other opinions and point of view — thereby engendering trust in relations between people" (Stata 1994, p. 38 f). These definitions imply that openness encourages new and innovative ideas and risk taking, increases a cross-fertilization and -functional support of ideas and signals to employees that they are valued, which in turn encourages those employees to care about innovation for their firm's competitive advantage (cf. Hurley & Hult 1998, p. 46 f.).

Von Kortzfleisch (2004) has investigated procedure models for the development of ICT between closed and open design patterns. He highlights the advantage of open procedure models such as prototyping: development processes must be passed through multiplication, which allows feedback from colleagues or customers again and again, allows for better adaptability, reduces the risk of faulty conceptualizations and enables creativity (cf. von Kortzfleisch 2004, p. 375 f.). These advantages can be transferred to open knowledge transfers.

Bettinger (1989) has linked firms' openness to weaknesses and strengths of firm culture in terms of communication: "In companies with weak cultures, the emphasis is almost exclusively on formalized downward communication. In strong culture companies, communication tends to be much more open, spontaneous, and informal" (Bettinger 1989, p. 40). Connecting openness with the attitude toward change, *Bettinger* (1989) has sensibilized the following point: "As it relates to 'problem solving' a positive leadership attitude toward change constantly reinforces the expectation of discovering creative solutions. In fact, 'problems' are more likely to be perceived as opportunities with the only real impediments being those that are self-imposed" (Bettinger 1989, p. 38 f.). This type of organizational culture is very important for a firm, because innovation is encouraged and there is a constant challenge to find better processes for task performance (cf. Bettinger 1989, p. 38 f.).

3.2.2.4   Autonomy

Autonomy has been defined by *Hackman & Oldham* (1976) as the "degree to which the job provides substantial freedom, independence, and discretion to the individual in scheduling the work and in determining the procedures to be used in carrying it out" (Hackman & Oldham 1976, p. 258). *Janz & Prasarnphanich* (2003) have supported this definition in the following statement: "Autonomy, equivalently referred to as 'self-direction,' 'empowerment,' or 'self-management,' is the extent to which an individual or group of individuals has the freedom, independence, and direction to determine what actions are required and how best to execute them" (Janz & Prasarnphanich 2003, p. 359)[62].

Autonomy arises from a culture that is designed to support and encourage the recognition, sharing, communication and creation of new knowledge (cf. Pemberton & Stonehouse 2002, p. 87). *Davenport et al.* (1997) have dealt with the issue of fit between organizational culture, knowledge management initiatives and autonomy, giving the following example: "At Hewlett-Packard, knowledge management projects are popping up all around the firm, but they are highly decentralized. Top management realizes the firm's culture of highly autonomous business units would not support a coordinated, top-down project at the corporate level, or even a corporate-level senior knowledge executive. Projects that don't fit the culture most likely won't thrive, so management needs to align its approach with its existing culture, or be prepared for a long-term culture change effort" (Davenport et al. 1997, p. 16).

Schein (1990) has connected autonomy with the decision process: "For example, as this company grows, the decision process may prove to be too slow, the individual autonomy that members are expected to exercise may become destructive and have to be replaced by more disciplined behavior, and the notion of a family may break down because too many people no longer know each other personally" (Schein 1990, p. 113). Thus, the growth of a company can be stunted if decisions are made too slowly because of too little autonomy; if a company grows too slowly, it can experience negative consequences with regard to its organizational culture.

---

[62] *Janz & Prasarnphanich* (2003, p. 359), referencing *Henderson & Lee* (1992), *Manz* (1992) and *Manz & Sims* (1980).

3.2.2.5 Learning Receptivity

Allee (2001) has been concerned with learning in companies and states that knowledge is embodied in people: "It's impossible to talk about knowledge without addressing the way people work together, learn together, and grow in knowledge individually and collectively" (Allee 2001, p. 3). Therefore, companies that are serious about knowledge must foster an environment and culture that support learning and employees' learning receptivity (cf. Allee 2001, p. 3). An important influencing factor for employees' learning receptivity is to facilitate learning by using knowledge sources who have similar training and background characteristics to those of the employees: "One of the most important ways that people learn new ideas is by associating those ideas with what they already know. As a result, people find it easier to absorb new ideas in areas in which they have some expertise and find it more difficult to absorb new ideas outside of their immediate area of expertise. An implication is that it is easier for knowledge to transfer from the source to a recipient when the source and the recipient have knowledge in common. Consequently, knowledge is more likely to be transferred between people with similar training and background characteristics" (Reagans & Mcevily 2003, p. 243). *Autio et al.* (2000) have added an interesting point related to learning receptivity. He has highlighted that the assimilation of new knowledge involves not only learning the new knowledge but also unlearning the old knowledge (cf. Autio et al. 2000, p. 911)[63]. This emphasizes that learning receptivity concerns not only new knowledge but also old knowledge and therefore, existing routines, learning processes, etc.

3.2.2.6 Care

To care for someone means "to help her to learn, to help her to increase her awareness of important events and their consequences, and to help nurture her personal knowledge creation while sharing her insights" (von Krogh 1998, p.

---

[63] *Autio et al.* (2000, p. 911), referencing *Bettis & Prahalad* (1995) and *Nonaka* (1994).

137). Therefore, relationships are built up and the behavior and interplay of those relationships are characterized by active empathy (cf. von Krogh 1998, p. 137)[64]. This active empathy is fostered by care, making it possible to assess and understand what the other person needs (cf. von Krogh 1998, p. 137). This assessment and understanding is very important for utilizing external knowledge. "Care also translates into real help" (von Krogh 1998, p. 138), e.g., when knowledge is shared with colleagues. Furthermore, care both fosters lenient judgment among members during knowledge-sharing and is visible in the courage that members exhibit towards each other (cf. von Krogh 1998, p. 138). This courage is very important for empowering employees to give opinions, propose ideas and give feedback (cf. Sollberger 2006, p. 127).

After having identified and described the parameters of the model, the construct of absorptive capacity based on *Zahra & George* (2002) and the construct of organizational culture, in the following sections the relationships among the capabilities of absorptive capacity and the dimensions of culture are in the foreground.

3.3 The Relationship between Organizational Culture and the Capabilities of Absorptive Capacity

For reviewing the research problem and achieving the objective of this thesis, hypotheses about the relationships between the dimensions of organizational culture and acquisition, assimilation, transformation and exploitation are discussed from section 3.3.1 through section 3.3.4. The hypotheses are derived from theoretical statements on absorptive capacity and organizational culture that have been made in the literature.

---

[64] *Von Krogh* (1998, p. 137), referencing *Gaylin* (1976) and *Mayeroff* (1971).

The discussion of those hypotheses is expanded to aspects of the theories that inform absorptive capacity: organizational learning, innovation, dynamic capabilities, knowledge-based view, managerial cognition and coevolution. Therefore, each discussion of the relationship between a dimension of organizational culture and a capability of absorptive capacity refers to the theory that informs it[65]. First, section 3.3.1 discusses the relationship among trust, collaboration, openness, autonomy, learning receptivity, care and acquisition.

3.3.1 The Relationship Between Organizational Culture and the Acquisition Capability

3.3.1.1 The Relationship Between Trust and the Acquisition Capability

Trust is very important for a firm's capacity to acquire external knowledge because trust between a sender and receiver is a key factor of influence for sharing knowledge. The reason that trust between a sender and receiver is important is that trustworthiness is a factor that influences knowledge-sharing decisions (cf. Andrews & Delahaye 2000, p. 797). If trust is high, people are more willing to engage in knowledge sharing (cf. Nahapiet & Ghoshal 1998, p. 254). *Nelson & Cooprider* (1996) have motivated this assumption in stating that by creating common expectations and reducing individual dissonance-inducing fears among employees, trust brings people closer together, and the attainment of trust leads to knowledge acquisition (cf. Nelson & Cooprider 1996, p. 413).

From a *knowledge-based view*, the acquisition of external knowledge through sharing is the key to developing and increasing a receiver's knowledge base and therefore its central competitive advantage (cf. Volberda et al. 2010, p. 935). Trust is inevitable for the acquisition of external knowledge and therefore, complementing the receiver's knowledge base: If a sender is unwilling to share

---

[65] The theories are italicized.

knowledge, a receiver cannot acquire knowledge. If a receiver can acquire knowledge, he can directly, positively affect his firm's acquisition capability, when he acts from a *managerial cognition* point of view (such as that of an information worker) and provides the knowledge to those potential adopters in the firm who he trusts (cf. Lenox & King 2004, p. 332).

Based on the literature's theoretical statements about absorptive capacity (especially on acquisition) and organizational culture (especially on trust) and based on the knowledge-based view and managerial cognition as theories informing absorptive capacity, it is assumed that trust is positively related to the acquisition capability.

Hypothesis H1: Trust is positively related to the acquisition capability.

3.3.1.2 The Relationship Between Collaboration and the Acquisition Capability

It is not only trust that is positively related to the acquisition capability. It is important to consider that the acquisition of external knowledge is more easily accomplished in organizations that have enhanced collaboration in an open *innovation* atmosphere where knowledge flows into and out of the company because organizational boundaries to innovation are porous (cf. Molina & Lloréns-Montes 2006, p. 266; cf. Morris, Kuratko & Covin 2008, p. 97). This enhanced collaboration during the acquisition of external knowledge is very important for a firm's competitive advantage because, following the *knowledge-based view*, the central competitive dimension of what firms know how to do is to collaborate and share knowledge within the organizational context (cf. Kogut & Zander 1992, p. 384).

Within close relationships between collaborative people, knowledge can be shared easily. This assumption is supported by the results of an investigation by *Reagans & McEvily* (2003). Their results have indicated that both social cohesion, which is the extent to which a relationship is surrounded by connections to third parties, and range, which is the extent to which network connections span organizational or social boundaries, ease knowledge transfer (cf. Reagans & McEvily 2003, p. 245 ff.). *Reagans & McEvily* (2003) have explained these results by stating that a strong social cohesion increases the willingness of individuals to devote time and effort to assisting others (cf. Reagans & McEvily 2003, p. 245).

Companies that employ individuals who have an increased willingness to devote their time and effort to assisting others can potentially offer more and better outputs by following an open *innovation* model and accessing external knowledge. This is because if people have different knowledge and more ideas to choose from for their innovations, the cost, quality and speed of acquisition and innovation improve (cf. Rigby & Zook 2002, p. 82).

From a *managerial cognition* perspective, managers can support collaboration during acquisition by directly providing information (cf. Lenox & King 2004, p. 332 ff.): "People acquire knowledge through practice, such as the practice of transferring knowledge within the group, to carry out a common task, which legitimates and brings about the acquisition of the knowledge needed for this transfer to take place, both within the group and also with other groups" (Molina & Lloréns-Montes 2006, p. 266)[66]. This statement supports the positive relationship between collaboration and the acquisition capability because it notes that people acquire knowledge by carrying out common tasks, through collaborating.

Based on the literature's theoretical statements about absorptive capacity (especially on acquisition) and organizational culture (especially on collaboration), and based on innovation, the knowledge-based view and managerial

---

[66] *Molina & Lloréns-Montes* (2006, p. 266), referencing *Orlikowski* (2002).

3 A Model of External Knowledge Absorption   91

cognition as theories informing absorptive capacity, it is assumed that collaboration is positively related to the acquisition capability.

> *Hypothesis H2: Collaboration is positively related to the acquisition capability.*

3.3.1.3 The Relationship Between Openness and the Acquisition Capability

Pemberton & Stonehouse (2002) have highlighted that when the acquisition of knowledge is accompanied by openness, such as during the acquisition of external knowledge, knowledge and ideas flow into and out of a company because organizational boundaries to innovation are porous (cf. Pemberton & Stonehouse 2002, p. 79). *Knudsen et al.* (2001) has emphasized the importance of openness for the absorption and especially the acquisition of external knowledge for open *innovation* as follows: "*Openness* towards knowledge sharing is a precondition for knowledge *access*; This access broadens the *absorptive capacity* as well as creates new knowledge; Absorptive capacity opens up for the *utilization* of knowledge [;] These steps each have a positive impact on the *innovative performance* of the firm" (Knudsen et al. 2001, p. 2)[67]. Therefore, according to *Knudsen et al.* (2001), openness towards knowledge sharing is a precondition for knowledge access and therefore, the acquisition of external knowledge to foster the innovation that is most strategically important to the companies' resources from a *knowledge-based view* (cf. Grant 1996a, p. 110). Other authors have supported the notion that openness is important for the absorption of knowledge and in particular, the identification and acquisition of external knowledge. For example, *Cohen & Levinthal* (1994) have suggested from a *coevolution*ary perspective that it is wrong for managers to believe that in the face of emerging technologies, uncertainty is best resolved by waiting

---

[67] The style of English was changed from British to American.

passively for clearer signals from the environment. Instead, managers must invest in their knowledge and absorptive capacity to preempt environmental changes. Davenport & Prusak (1998) have supported the necessity of a sufficient absorptive capacity and have suggested hiring for openness to ideas as a possible solution for a lack of absorptive capacity (cf. Davenport & Prusak 1998, p. 97). This solution follows an open *innovation* model (cf. Chesbrough 2006).

The acquisition of new ideas within the knowledge structure is considered as an important factor influencing *managerial cognition* (cf. Lenox & King 2004, p. 332 ff.). Managers that are assumed to be information workers themselves may have a lack of knowledge necessary to efficiently discover and acquire new practices as a critical element of sustained competitive advantage (cf. Lenox & King 2004, p. 332). Thus, more knowledge acquired from external sources may lead to a higher degree of *managerial cognition*.

To summarize, the capability of acquisition is improved through openness because in particular, the acquisition of knowledge is accelerated and enhanced by the sharing of knowledge (cf. Pemberton & Stonehouse 2002, p. 79) accompanied by an openness that encourages the accessibility of information, open communication, the sharing of problems, discussion, etc. during *organizational learning* (cf. Appelbaum & Gallagher 2000, p. 50). With regard to organizational learning, it is important that there is not only an openness to external knowledge but also to external training and background characteristics because the more similar the training and background characteristics of employees, the higher the probability that knowledge will be transferred between them. It is very important to keep in mind that a firm's potential to learn is determined by prior related knowledge because the higher the prior related knowledge of a firm's employees, the higher the probability that they will acquire knowledge.

Based on the literature's theoretical statements on absorptive capacity (especially on acquisition) and organizational culture (especially on openness), and based on organizational learning, innovation, the knowledge-based view, managerial cognition and coevolution as theories informing absorptive capacity, it is assumed that openness is positively related to the acquisition capability.

> *Hypothesis H3: Openness is positively related to the acquisition capability.*

3.3.1.4 The Relationship Between Autonomy and the Acquisition Capability

Nonaka (1994) has investigated the influence of autonomy on the absorption of knowledge, which starts with the acquisition of knowledge: "Individual autonomy widens the possibility that individuals will motivate themselves to form new knowledge. Self-motivation based on deep emotions, for example, in the poet's creation of new expressions, serves as a driving force for the creation of metaphors. A sense of purpose and autonomy becomes important as an organizational context. Purpose serves as the basis of conceptualization. Autonomy gives individuals freedom to absorb knowledge" (Nonaka 1994, p. 18). This statement supports the assumption that autonomy is positively related to the acquisition capability because it shows that autonomy fosters employees' absorption of knowledge and the acquisition of external knowledge. Knowledge acquisition is, from a *knowledge-based view*, the most strategically important of a firm's resources, and external knowledge that has the potential to complement a firm's existing knowledge base is highly desired. Therefore, it is very important that employees become autonomous with the acquisition of external knowledge.

When employees become autonomous with the acquisition of external knowledge, the effort expended in knowledge acquisition routines includes attributes such intensity and speed, which can influence absorptive capacity (cf. Zahra & George 2002, p. 189). The intensity and speed of a firm's efforts to identify and gather knowledge can determine the quality of the acquisition capability (cf. Zahra & George 2002, p. 189). Therefore, a firm's ability to achieve speed is limited by the difficulties and efforts of both short learning cycles and the assembly of the resources needed to build *dynamic capabilities* that are inevitable for the acquisition capacity, which represent the firm's ability to integrate, build and reconfigure competences (cf. Clark & Fujimoto 1991, p. 206 f.). Autonomy can help to shorten *organizational learning* cycles and to assemble the resources needed to build acquisition capability. The potential of a firm to learn and change its learning cycles is determined by prior related knowledge because the higher the level of employees' prior related knowledge, the higher the probability that they will acquire knowledge. When employees become autonomous with the acquisition of external knowledge they indirectly increase their prior related knowledge and therefore shorten organizational learning cycles.

Investigating autonomy from a *managerial cognition* perspective, the "most fundamental challenge faced by managers, however, is that their information worlds are extremely complex, ambiguous, and munificent [...]. Somehow they must see their way through what may be a bewildering flow of information to make decisions and solve problems" (Walsh 1995, p. 280)[68]. Autonomy can facilitate decision-making and solve the problem that speed is limited by difficulties due to complex, ambiguous, and munificent knowledge.

Based on the literature's theoretical statements on absorptive capacity (especially on acquisition) and organizational culture (especially on autonomy), and based on organizational learning, dynamic capabilities, the knowledge-based view and managerial cognition as theories informing absorptive capacity, it is assumed that autonomy is positively related to the acquisition capability.

Hypothesis H4: Autonomy is positively related to the acquisition capability.

---

[68] *Walsh* (1995, p. 280), referencing *Mason & Mitroff* (1981), *Mintzberg, Raisinghani & Théorêt* (1976), *Schwenk* (1984), and *Starbuck & Milliken* (1988).

3.3.1.5 The Relationship Between Learning Receptivity and the Acquisition Capability

Investigating the relationship between learning receptivity as a further dimension of organizational culture and acquisition, it is important to consider that *organizational learning* is defined as "the acquisition of new knowledge by people who are able and willing to apply that knowledge in making decisions or influencing others" (Choi 2002, p. 52)[69]. This definition highlights that learning receptivity is a precondition to identifying and acquiring new knowledge. *Minbaeva et al.* (2003) have supported this assumption, stating that from a *managerial cognition* perspective, absorptive capacity is enhanced by an individual ability to reduce complexity using both mental maps developed through experience in the core business and individual motivations (cf. Minbaeva et al. 2003, p. 597). More knowledge acquired from external sources may lead to a higher degree of managerial awareness and cognition. Furthermore, managers can directly, positively affect the absorptive capacity of their firms when they provide knowledge to potential adopters within those firms (cf. Lenox & King 2004, p. 332).

Absorptive capacity is important for creating dynamic capabilities, and therefore companies need learning receptivity to integrate, build and reconfigure internal and external knowledge during the acquisition of knowledge to address rapidly changing environments to generate competitive advantage (cf. Teece et al. 1997, p. 516). Furthermore, firms with a higher level of absorptive capacity tend to be more proactive and therefore are better at anticipating the emergence of valuable developments through *coevolution* because proactive firm behavior is very important in turbulent environments (cf. van den Bosch et al. 1999, p. 552).

Experience is the key for absorptive capacity and the acquisition of knowledge because it is extremely difficult for people to share each other's

---

[69] *Choi* (2002) summarizes the results of *Miller* (1996).

thinking processes and transfer knowledge abstracted from the embedded emotions and nuanced contexts that are associated with shared experiences (cf. Nonaka 1994, p. 19). *Zahra & George* (2002) have highlighted that "experience is the product of environmental scanning [...], interaction with customers [...], and alliances with other firms" (Zahra & George 2002, p. 193)[70]. Furthermore experience can be gained from learning receptivity to learning by doing (cf. Levitt & March 1988, p. 131). Therefore, learning receptivity is necessary for experience, which is the key for the acquisition of knowledge. From a *knowledge-based view*, learning receptivity can support a firm's ability of firms to create and transfer knowledge within its organizational context and therefore create a competitive advantage.

With regard to the three levels of organizational culture, *Schein* (1990) has given the following explanation: "Once a group has learned to hold common assumptions, the resulting automatic patterns of perceiving, thinking, feeling, and behaving provide meaning, stability, and comfort; the anxiety that results from the inability to understand or predict events happening around the group is reduced by the shared learning" (Schein 1990, p. 111). He has annotated that the "strength and tenacity of culture derive, in part, from this anxiety reduction function" (Schein 1990, p. 111). This reduction of anxiety is very important: "When employees feel psychologically safe, they engage in learning behavior — they ask questions, seek feedback, experiment, reflect on results, and discuss errors or unexpected outcomes openly" (Chatman & Cha 2003, p. 25). *Chatman & Cha* (2003) have picked up on *Edmondson* (1999), who has made the following point: "Team psychological safety should facilitate learning behavior in work teams because it alleviates excessive concern about others' reactions to actions that have the potential for embarrassment or threat, which learning behaviors often have" (Edmondson 1999, p. 355). He has substantiated this statement with the following example: "For example, team members may be unwill-

---

[70] *Zahra & George* (2002, p. 193), referencing *Fahey* (1999), *Lane & Lubatkin* (1998) and *Nonaka & Takeuchi* (1995).

3 A Model of External Knowledge Absorption

ing to bring up errors that could help the team make subsequent changes because they are concerned about being viewed as incompetence, which allows them to ignore or discount the negative consequences of their silence for team performance. In contrast, if they respect and feel respected by other team members and feel confident that team members will not hold the error against them, the benefits of speaking up are likely to be given more weight" (Edmondson 1999, p. 355).

Employees will feel safe and learn whether there is an established organizational culture that shows them how to address ideas and errors, which in turn has an impact on the perception of the consequences of errors. This perception affects employees' willingness to report errors and is important for employees' ability to identify and analyze problems and to acquire knowledge to solve those problems. Furthermore, the import and export of ideas helps companies to clarify what they do best (Rigby & Zook 2002, p. 84) because companies often think that their core business is broader than it really is, and the market knowledge gained by open *innovation* helps them to discover where they are stronger and weaker in reality than previously assumed. Firms can and should acquire and use both external and internal knowledge to accelerate internal innovation and to expand the market for the external use of innovation. The basic prerequisite for the acquisition and use of external knowledge is learning receptivity.

Based on the literature's theoretical statements on absorptive capacity (especially on acquisition) and organizational culture (especially on learning receptivity), and based on organizational learning, innovation, the knowledge-based view, managerial cognition and coevolution as theories informing absorptive capacity, it is assumed that learning receptivity is positively related to the acquisition capability.

Hypothesis H5: Learning receptivity is positively related to the acquisition capability.

3.3.1.6 The Relationship Between Care and the Acquisition Capability

As the sixth dimension of organizational culture, the relationship between care and acquisition must be investigated. Because care is necessary to increase the awareness of important knowledge and to help nurture personal knowledge creation while sharing insights, care supports the acquisition of external knowledge (cf. von Krogh 1998, p. 137). Furthermore, from a *knowledge-based view*, care that complements a firm's existing knowledge base has a positive effect on the acquisition of external knowledge (cf. Lichtenthaler 2009, p. 822 ff.). Care gives rise to active empathy so that people can assess and understand what others need (cf. von Krogh 1998, p. 137). According to *managerial cognition*, this assessment and understanding is very important for utilizing external knowledge and adapting the organizational knowledge structure, which is built out of a social process (cf. Lyles & Schwenk 1992, p. 170 f.).

Based on the literature's theoretical statements on absorptive capacity (especially on acquisition) and organizational culture (especially on care), and based on the knowledge-based view and managerial cognition as theories informing absorptive capacity, it is assumed that care is positively related to the acquisition capability.

> *Hypothesis H6: Care is positively related to the acquisition capability.*

3.3.1.7 Summary of the Relationship Between Organizational Culture and the Acquisition Capability

In sum, the six dimensions of organizational culture are positively related to the acquisition capability.

3 A Model of External Knowledge Absorption 99

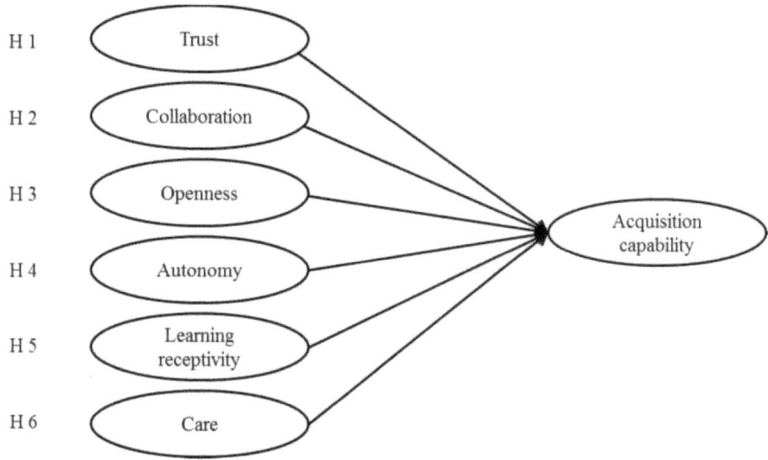

Figure 14 Research Model of the Acquisition Capability[71]

Organizational Theories Informing the Acquisition Capability

With regard to the organizational theories informing the acquisition capability, Table 8 shows:

- **Organizational learning** informs three dimensions of organizational culture (openness, autonomy and learning receptivity),
- **Innovation** informs three dimensions of organizational culture (collaboration, openness and learning receptivity),
- **Dynamic capabilities** inform one dimension of organizational culture (autonomy),
- The **knowledge-based view** informs all six dimensions of organizational culture (trust, collaboration, openness, autonomy, learning receptivity and care),

---

[71] Author's own figure.

- **Managerial cognition** informs all six dimensions of organizational culture (trust, collaboration, openness, autonomy, learning receptivity and care) and
- **Coevolution** informs two dimensions of organizational culture (openness and learning receptivity).

The acquisition capability is mostly informed by the knowledge-based view and managerial cognition, followed by organizational learning and innovation. The remaining theories — coevolution and dynamic capabilities — play only a small role.

With regard to the most important theories informing absorptive capacity, the knowledge-based view is so important for the acquisition capability because detailed aspects of the acquisition capability include the identification of new knowledge and the acquisition of knowledge relevant to a company's operations[72]. From a knowledge-based view, the main determinant of competitive advantage is a firm's ability to identify and absorb knowledge and to develop and increase its knowledge base. Knowledge is the most strategically important of a firm's resources and firms greatly desire external knowledge that has the potential to complement their existing knowledge bases (cf. Grant 1996a, p. 110). Therefore, the knowledge-based view plays an important role for the acquisition capability because the acquisition capability refers to a firm's capability to identify and acquire knowledge relevant to a company's operations from external knowledge sources (cf. Flatten et al. 2011a, p. 100, cf. Flatten et al. 2011b, p. 138, cf. Zahra & George 2002, p. 189).

Managerial cognition is very important for the acquisition capability because detailed aspects of acquisition capability include prior knowledge as a prerequisite and learning from partners. Managers are 'information workers' and build a basis for acquisition because they are the owners of prior knowledge (cf. McCall & Kaplan 1985). Furthermore, managerial cognition theory implies

---

[72] The detailed aspects of the acquisition capability are explained in section 3.2.1.1.

that the effectiveness of the provision of managerial information is contingent on the degree of the recipient of the knowledge. This highlights the importance of the ability to learn from partners.

Table 8 Theories Informing the Acquisition Capability[73]

	Organizational Learning	Innovation	Dynamic Capabilities	Knowledge-Based View	Managerial Cognition	Coevolution
Trust				X	X	
Collaboration		X		X	X	
Openness	X	X		X	X	X
Autonomy	X		X	X	X	
Learning Receptivity	X	X		X	X	X
Care				X	X	
Sum	3	3	1	6	6	2

3.3.2 The Relationship Between Organizational Culture and the Assimilation Capability

3.3.2.1 The Relationship Between Trust and the Assimilation Capability

Trust is very important with regard to the assimilation of external knowledge because it is a frame of reference that allows persons to conserve cognitive resources, which are, according to the *knowledge-based view*, the most valuable resource from which a firm might derive its competitive advantage from (cf. McEvily et al. 2003, p. 93). The reason is that trust represents an expectation that another person will act in accordance with one's own desires (cf. McEvily

---

[73] Author's own table.

et al. 2003, p. 93)[74]. When trust is high, the overall knowledge exchange and the likelihood that the resources acquired from other persons will be assimilated increase (cf. Abrams, Cross, Lesser & Levin 2003, p. 65).

With regard to acquired resources, to assimilate knowledge the recipient must have trust in the sender about its accuracy. In addition, to be willing to provide the knowledge the sender must trust the recipient. He must trust that the recipient will address the knowledge responsibly and be certain that the recipient will not abuse it, e.g., by claiming that it is the creation of his personal contribution (cf. Kunz 2010, p. 35). Therefore, trust in the skills and knowledge of the sender is important for both the assimilation of external knowledge and the development of insights and relationships that are necessary to assimilate it (cf. Kunz 2010, p. 35).

Because trust represents a positive assumption about knowledge-senders' motives and intentions, it allows receivers to economize on knowledge processing (cf. McEvily et al. 2003, p. 92 f.). From a *managerial cognition* perspective, trust makes decision-making more efficient by simplifying the acquisition and interpretation of knowledge that corresponds to the assimilation of knowledge (cf. McEvily et al. 2003, p. 93).

Based on the literature's theoretical statements on absorptive capacity (especially on assimilation) and organizational culture (especially on trust), and based on the knowledge-based view and managerial cognition as theories informing absorptive capacity, it is assumed that trust is positively related to the assimilation capability.

Hypothesis H7: Trust is positively related to the assimilation capability.

---

[74] *McEvily et al.* (2003, p. 93), referencing *Gambetta* (1988) and *Uzzi* (1997).

3.3.2.2 The Relationship Between Collaboration and the Assimilation Capability

Interaction within collaboration is essential for developing and understanding knowledge (cf. Sollberger 2006, p. 123), and therefore understanding of knowledge is inevitable for the assimilation of knowledge. In assimilation, externally acquired knowledge must be analyzed, interpreted and, especially, understood (cf. Zahra & George 2002, p. 189). The degree of interaction within a collaboration is dependent on the type of cooperation, as *Probst, Raub & Romhardt* (2010) showed when they compared, e.g., mergers, joint ventures, strategic networks, and occasionally information related to their degree of cooperation, their access to a partner's knowledge base and capital expenditures (cf. Probst et al. 2010, p. 101).

O'Dell & Grayson (1999) have discovered that organizations with a culture that is based on collaboration are much more accepting of knowledge-sharing (cf. O'Dell & Grayson 1999, p. 13 f.) that is very important for both the acquisition and the assimilation of knowledge. The open *innovation* paradigm assumes that firms can and should use both external and internal knowledge, and internal and external paths to market, to advance their technology (cf. Chesbrough 2006, p. xxiv). Therefore, the just-mentioned differences in cultural values within organizations can lead to divergent organizational outcomes from knowledge management and intended absorption (cf. Alavi, Kayworth & Leidner 2006, p. 217). As a consequence, it is very important to create a collaborative environment for the assimilation of externally acquired knowledge, particularly because acquired knowledge can embody heuristics that differ significantly from those used by a company that can result in delayed comprehension of the knowledge during *organizational learning* (cf. Zahra & George 2002, p. 189)[75]. *Kogut & Zander* (1992) have supported the assumption that the ability to comprehend knowledge within an organizational context is a competi-

---

[75] *Zahra & George* (2002, p. 189), referencing *Leonard-Barton* (1995).

tive advantage, which is the focus of the *knowledge-based view* (cf. Kogut & Zander 1992, p. 384).

Based on the literature's theoretical statements on absorptive capacity (especially on assimilation) and organizational culture (especially on collaboration), and based on organizational learning, innovation and the knowledge-based view as theories informing absorptive capacity, it is assumed that collaboration is positively related to the assimilation capability.

> *Hypothesis H8: Collaboration is positively related to the assimilation capability.*

3.3.2.3 The Relationship Between Openness and the Assimilation Capability

Openness can be understood in terms of overall perceived openness as both the degree to which people work closely together on a common task and the degree to which people perceive that others withhold their knowledge (cf. Wathne et al. 1999, p. 61). Openness to *innovation* is essential for the assimilation of external knowledge because openness enables knowledge-sharing between different organizational units and hierarchical levels (cf. Sollberger 2002, p. 124). Therefore, a positive relationship between openness and the assimilation capability can be assumed. *Davenport & Prusak* (1998) have supported this statement, explaining that the exchange of knowledge in an open atmosphere enables employees at all levels to understand what is happening in a firm (cf. Davenport & Prusak 1998, p. 49). This understanding of a firm's routines and processes is essential for the assimilation of external knowledge, because as already mentioned, according to the *knowledge-based view* creating and transferring knowledge in a company creates a competitive advantage.

Based on the literature's theoretical statements on absorptive capacity (especially on assimilation) and organizational culture (especially on openness), and based on innovation and the knowledge-based view as theories informing absorptive capacity, it is assumed that openness is positively related to the assimilation capability.

> Hypothesis H9: Openness is positively related to the assimilation capability.

3.3.2.4 The Relationship Between Autonomy and the Assimilation Capability

Autonomy arises from a culture that is, in the *knowledge-based view*, designed to support and encourage employees to recognize, share, communicate and create new knowledge as their firm's most important resource (cf. Pemberton & Stonehouse 2002, p. 87). To assimilate and use external knowledge, SMEs must have prior related knowledge (e.g., basic skills, a shared language and knowledge of scientific or technological developments in a given field; cf. Cohen & Levinthal 1990, p. 128 f.) because the ability to evaluate and utilize external knowledge is largely a function of prior related knowledge. Therefore, knowledge must be recognized, shared and communicated to foster the assimilation of external knowledge. With regard to this recognition, sharing and communication, *Molina & Lloréns-Montes* (2006) have highlighted that "in decision making, it is normal to have to bring together a large amount of knowledge that is scattered throughout the firm, especially when the individuals have specific complex knowledge gained from their prior training [...]. This means that the decision-making authority is obliged to look for the necessary knowledge within the organization and attempt to transfer it to the work groups where it is needed at a given time" (Molina & Lloréns-Montes 2006, p. 267). Therefore, the autonomy to make decisions can enhance the assimilation of external knowledge because the prior related knowledge is inevitable.

Based on the literature's theoretical statements on absorptive capacity (especially on assimilation) and organizational culture (especially on autonomy), and based on the knowledge-based view as a theory informing absorptive capacity, it is assumed that autonomy is positively related to the assimilation capability.

> *Hypothesis H10: Autonomy is positively related to the assimilation capability.*

3.3.2.5 The Relationship Between Learning Receptivity and the Assimilation Capability

Learning receptivity is not only positively related to the acquisition capability. Hurley & Hult (1998) have noted that *organizational learning* has "an emphasis on individual learning and development infuses the organization with new ideas [...] [,] enhances the capacity to understand new ideas [...] [,] enhances creativity and the ability to notice novel opportunities [...] [and] aids in implementation by improving problem solving" (Hurley & Hult 1998, p. 46; own formatting). Therefore, learning receptivity is also positively related to the assimilation capability.

During the assimilation, the development of routines and processes that allow analyzing, processing, interpreting, and understanding of the previously mentioned new ideas to create *dynamic capabilities* is promoted by creating a shared belief that it is safe for employees to take interpersonal risks. Following the *knowledge-based view* once again, the behavior of employees — asking questions, seeking feedback, experimenting, reflecting on results, etc. — is inevitable for the development of a firm's capability to develop and refine those routines that facilitate combining existing knowledge with the acquired and assimilated knowledge for future use to gain competitive advantage (cf. Flatten et al. 2011a, p. 100, cf. Zahra & George 2002, p. 190). Managers and their *managerial cognition* can help to assimilate huge and complex amounts of knowledge because they can reduce complexity by using mental maps developed through experience in the core business, which are sometimes inappropriately applied to other businesses (cf. Prahalad & Bettis 1986, p. 485).

In addition to managerial cognition, R&D can help to assimilate knowledge. *Cohen & Levinthal* (1989) have investigated absorptive capacity, have assumed that R&D generates new knowledge and enhances the assimilation capability of existing knowledge, and have considered the implications of this dual role of R&D for the firm's incentive to invest in R&D (cf. Cohen & Levinthal 1989, p. 569). They have argued that recognition of the role of R&D as an enhancer of a firm's ability to assimilate existing information suggests "that the ease and character of learning within an industry will both affect R&D spending and condition the influence of appropriability and technological opportunity conditions on R&D" (Cohen & Levinthal 1989, p. 569).

Based on the literature's theoretical statements on absorptive capacity (especially on assimilation) and organizational culture (especially on learning receptivity), and based on organizational learning, dynamic capabilities, the knowledge-based view and managerial cognition as theories informing absorptive capacity, it is assumed that learning receptivity is positively related to the assimilation capability.

> *Hypothesis H11: Learning receptivity is positively related to the assimilation capability.*

3.3.2.6 The Relationship Between Care and the Assimilation Capability

In addition to the relationship between learning receptivity and assimilation, the relationship between care and assimilation must be investigated. *Lyles & Schwenk* (1992) have noted the increased interest on the part of the strategic management research in the cognitions of top management teams and have discussed the existence and maintenance of organizational knowledge structures (cf. Lyles & Schwenk 1992, p. 170): "One of the crucial points about these and the difference between an individual's schema is that the organizational knowledge structure is built out of a social process" (cf. Lyles & Schwenk 1992, p. 170 f.). This social process, based on *managerial cognition*, involves caring for someone. *Von Krogh* (1998) has defines caring for a person as follows: "To care for someone is to help her to learn, to help her to increase her awareness of important events and their consequences, and to help nurture her personal knowledge creation while sharing her insights" (von Krogh 1998, p. 137)[76]. According to the *knowledge-based view*, this caring is inevitable for understanding knowledge obtained from external sources and for analyzing, interpreting and understanding that new knowledge.

Based on the literature's theoretical statements on absorptive capacity (especially on assimilation) and organizational culture (especially on care), and based on the knowledge-based view and managerial cognition as theories informing absorptive capacity, it is assumed that care is positively related to the assimilation capability.

> *Hypothesis H12: Care is positively related to the assimilation capability.*

3.3.2.7 Summary of the Relationship Between Organizational Culture and the Assimilation Capability

In summary, the six dimensions of organizational culture are positively related to the assimilation capability.

---

[76] *Sollberger* (2006, p. 127), referencing *von Krogh* (1998), which is also cited in this section 3.3.2.6.

3 A Model of External Knowledge Absorption 109

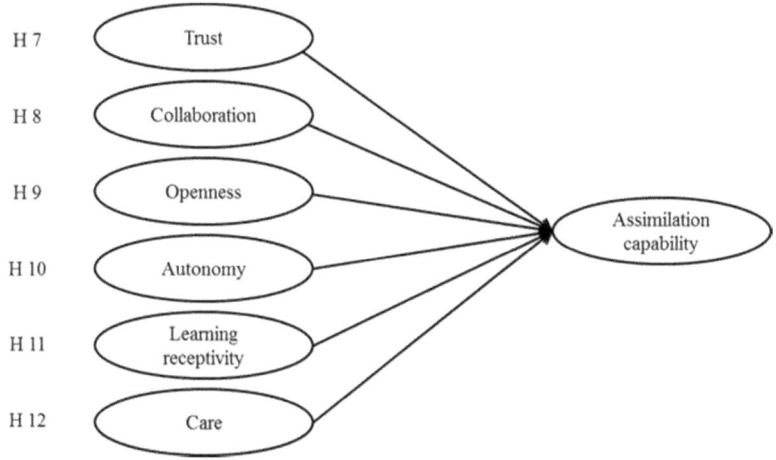

Figure 15 Research Model of the Assimilation Capability[77]

Organizational Theories Informing the Assimilation Capability

With regard to the organizational theories that inform the assimilation capability, Table 9 shows that

- **Organizational learning** informs two dimensions of organizational culture (collaboration and learning receptivity),
- **Innovation** informs two dimensions of organizational culture (collaboration and openness),
- **Dynamic capabilities** inform one dimension of organizational culture (learning receptivity),
- The **knowledge-based view** informs all six dimensions of organizational culture (trust, collaboration, openness, autonomy, learning receptivity and care),

---

[77] Author's own figure.

- **Managerial cognition** informs three dimensions of organizational culture (trust, learning receptivity and care) and
- **Coevolution** informs zero dimensions of organizational culture (/).

Table 9 Theories Informing the Assimilation Capability[78]

	Organizational Learning	Innovation	Dynamic Capabilities	Knowledge-Based View	Managerial Cognition	Coevolution
Trust				X	X	
Collaboration	X	X		X		
Openness		X		X		
Autonomy				X		
Learning Receptivity	X		X	X	X	
Care				X	X	
Sum	2	2	1	6	3	0

Like the acquisition capability, the assimilation capability is mostly informed by the knowledge-based view, followed by managerial cognition. Organizational learning, innovation and dynamic capabilities have less influence. Coevolution has no influence.

With regard to the most important theories that inform absorptive capacity, the knowledge-based view is the most important for the assimilation capability because one detailed aspect of the assimilation capability is developing routines and processes[79]. According to the knowledge-based view, "the central competitive dimension of what firms know how to do is to create and

---

[78] Author's own table.
[79] The detailed aspects of the assimilation capability are explained in section 3.2.1.2.

transfer knowledge efficiently within an organizational context" (Kogut & Zander 1992, p. 384). This means that developing routines and processes during the assimilation of external knowledge is the central competitive dimension.

Managerial cognition is also extremely important for the assimilation capability because a detailed aspect of the assimilation capability is analyzing, processing, interpreting and understanding new knowledge. Managers understand information through their own cognitive lenses (cf. Volberda et al. 2010, p. 933) and thereby reduce complexity using mental maps that are developed through experience in the core business and are sometimes inappropriately applied to other businesses (cf. Prahalad & Bettis 1986, p. 485). This enables them to develop new routines and processes. It is important for firms to develop new routines and processes because those routines and processes influence the locus of the future search for knowledge (cf. Zahra & George 2002, p. 193).

3.3.3 The Relationship Between Organizational Culture and the Transformation Capability

3.3.3.1 The Relationship Between Trust and the Transformation Capability

After knowledge is assimilated, it must be transformed to complement the firm's existing knowledge base (cf. Lichtenthaler 2009, p. 822 ff.). Because transformation refers to a firm's capability to develop and refine those routines that facilitate combining existing knowledge with acquired and assimilated knowledge for future use (cf. Flatten et al. 2011a, p. 100, cf. Zahra & George 2002, p. 190), it is important to take note of trust. The reason is that according to the *knowledge-based view*, trust between a sender and recipient of knowledge influences the recipient's ability to understand new knowledge to generate new resources for competitive advantage (cf. Lane et al. 2001, p. 1139).

From a *managerial cognition* perspective, the dominant general management logic through organizational forms, e.g., routines and processes, influences the transformation capability, particularly in environments where the understanding of new knowledge is complicated by complexity (cf. Van den Bosch et al. 1999, p. 560). Under these circumstances, trust is even more essential for the transformation of external knowledge because it enables firms to develop and refine routines and processes that facilitate combining new knowledge from the acquisition and assimilation with already-existing knowledge. These routines and processes are inevitable for the transformation of externally acquired knowledge (cf. Flatten et al. 2011a, p. 100, cf. Zahra & George 2002, p. 190). Only when a recipient understands the new knowledge is he able to add, delete and interpret knowledge in a different manner that transforms that knowledge.

Based on the literature's theoretical statements on absorptive capacity (especially on transformation) and organizational culture (especially on trust), and based on the knowledge-based view and managerial cognition as theories informing absorptive capacity, it is assumed that trust is positively related to the transformation capability.

Hypothesis H13: Trust is positively related to the transformation capability.

3.3.3.2 The Relationship Between Collaboration and the Transformation Capability

Because knowledge utilization, in the sense of merging, synthesizing and transforming acquired knowledge into higher-value products and services, is the aim of transformation, collaboration is necessary to foster the transformation of externally acquired knowledge (cf. Spender 1996, p. 46). Therefore, collaboration is viewed as a fundamental element for utilizing that external knowledge during transformation: "At the heart of it, knowledge utilization is a collaborative process" (Miles, Miles, Perrone & Edvinsson 1998, p. 286). *Miles et al.*

3 A Model of External Knowledge Absorption 113

(1998) have specified this in more detail from the *knowledge-based view*: "Whether between workers within a firm or in the transfer and utilization of knowledge between firms, knowledge-based approaches cannot succeed without effective collaboration" (Miles et al. 1998, p. 286). Therefore, an organization's transformation capability depends on links across its individual capabilities, which affect its *dynamic capabilities* (cf. Cohen & Levinthal 1990, p. 133).

The reason for the necessity of effective collaboration is that collaboration "reduces fear and increases openness and therefore encourages new ideas and risk taking [...] [,] nurtures and encourages innovative ideas [...][,] increases cross-fertilization and cross-functional support of ideas [...] and signals to employees that they are valued, which encourages them to care about innovation for the good of the organization" (Hurley & Hult 1998, p. 46 f.; own formatting). Therefore, collaboration can help people to develop and refine those routines that facilitate combining existing knowledge with acquired and assimilated knowledge for future use. With respect to *innovation*, it is important to keep in mind that the fundamental problem in innovation is not to find more new ideas but rather to establish refined and leveraged routines and processes so that organizations are both open to exploring new ideas and willing to back the most promising ones (cf. Denning 2005, p. 8). This problem is very important with regard to transformation. As has been noted by *Denning* (2005), the transformation of external knowledge builds on refining and leveraging routines and processes and therefore, ways of running an organization, which can be problematic for innovation.

Crossan, Lane & White (1999) have also referenced transformation, and have stated as follows: "When actions take place in concert with other members of a workgroup, the interpreting process quite naturally blends into the integrating process. Integrating entails the development of shared understanding and the taking of coordinated action by members of a workgroup. Actions that are deemed to be effective will be repeated. Initially, the workgroup informally makes this judgment about what actions should be replicated. Eventually, the workgroup may establish formal rules and procedures, and routines become embedded" (Crossan et al. 1999, p. 525). Because transformation refers to a firm's capability to develop and refine those routines that facilitate combining existing knowledge with acquired and assimilated knowledge for future use (cf. Flatten et al. 2011a. p. 100, cf. Zahra & George 2002, p. 190), this process of institutionalizing, which occurs during the actions that are taking place, is inevitable for the transformation capability. From a *managerial cognition* perspective, dominant logic can be helpful during the process of transforming external knowledge because it has an information-filter-like effect that is then incorporated into a company's strategy, systems, values, expectations and reinforced behavior during collaboration (cf. Bettis & Prahalad 1995, p. 7).

Based on the literature's theoretical statements on absorptive capacity (especially on transformation) and organizational culture (especially on collaboration), and based on innovation, dynamic capabilities, the knowledge-based view and managerial cognition as theories informing absorptive capacity, it is assumed that collaboration is positively related to the transformation capability.

> *Hypothesis H14: Collaboration is positively related to the transformation capability.*

3.3.3.3 The Relationship Between Openness and the Transformation Capability

With regard to the transformation capability, the so-called 'Not Invented Here' (NIH) syndrome plays an important role. NIH syndrome is defined as "the tendency of a project group of stable composition to believe it possesses a monopoly of knowledge of its field, which leads it to reject new ideas from outsiders to the likely detriment of its performance" (Katz & Allen 1982, p. 7). The syndrome therefore refers to a negative attitude towards openness, including both acquisition and assimilation, but especially transformation, because employees must accept knowledge and develop and refine those routines that facilitate the combining of existing knowledge with acquired and assimilated knowledge for future use.

Employees who believe that they possess a monopoly on knowledge in their area of specialization do not seriously consider the possibility that outsiders might produce important new knowledge relevant to the company that they can gain by open *innovation* (cf. Katz & Allen 1982, p. 7). *Laursen & Salter* (2006) have supported this assumption and have noted that NIH syndrome suggests that greater attention to external knowledge may confront employee resistance (cf. Laursen & Salter 2006, p. 137). Nevertheless — keeping NIH syndrome in mind[80] — openness in general is positively related to the acquisition capability. *Badaracco* (1991) has stated from a *knowledge-based view* that "openness is paramount in knowledge links because much of what the parties are trying to learn from each other, or create together, is so difficult to communicate. It is often embedded in a firm's practice and culture, and it can only be learned through working relationships that are not hampered by constraints" (Badaracco 1991, p. 142). Within these working relationships, the transformation capability shapes the entrepreneurial mindset and fosters entrepreneurial action: "It yields new insights, facilitates the recognition of opportunities, and, at the same time, alters the way the firm sees itself and is competitive landscape. It is in these varied activities that the genesis of new competencies can be found" (Zahra & George 2002, p. 190).

Based on the literature's theoretical statements on absorptive capacity (especially on transformation) and organizational culture (especially on openness), and based on innovation and the knowledge-based view as theories informing absorptive capacity, it is assumed that openness is positively related to the transformation capability.

---

[80] It will be explained in section 5.2.3 how to deal with NIH syndrome.

> *Hypothesis H15: Openness is positively related to the transformation capability.*

3.3.3.4 The Relationship Between Autonomy and the Transformation Capability

When investigating the relationship between autonomy and transformation, it is important to consider that autonomy is the "degree to which the job provides substantial freedom, independence, and discretion to the individual in scheduling the work and in determining the procedures to be used in carrying it out" (Hackman & Oldham 1976, p. 258). This freedom supports the transformation capability, because from a *knowledge-based view* employees must feel free to develop and refine those routines that facilitate combining existing knowledge with acquired and assimilated knowledge for future use. Thereby *Sollberger*'s (2006) point must be kept in mind: She raises awareness of the point that employees are reluctant to turn their knowledge and ideas into action if they have to be afraid of being sanctioned for errors (cf. Sollberger 2006, p. 125).

Firms can be viewed as governance structures enabling them to merge, synthesize and transform acquired knowledge into higher-value products and services (cf. Spender 1996, p. 46). In their work, *Molina & Lloréns-Montes* (2006) have considered the results of *Wruck & Jensen* (1994), when they highlighted that they "attribute the importance of distributing decision making among the workers to the fact that decisions should be made where the best information is to be found and, on many occasions, the necessary knowledge is held by the firm's workers themselves" (Molina & Lloréns-Montes 2006, p. 267). Therefore, autonomy enhances the chance to develop higher-value products and services using the best information and positively transforming external knowledge.

Based on the literature's theoretical statements on absorptive capacity (especially on transformation) and organizational culture (especially on autonomy), and based on the knowledge-based view as theory informing absorptive capacity, it is assumed that autonomy is positively related to the transformation capability.

> *Hypothesis H16: Autonomy is positively related to the transformation capability.*

3.3.3.5 The Relationship Between Learning Receptivity and the Transformation Capability

"A culture that is positively oriented toward knowledge is one where learning on and off the job is highly valued, and where hierarchy takes a back seat to experience, expertise, and rapid innovation" (Davenport et al. 1997, p. 15). Such a positive orientation toward knowledge is inevitable for the transformation of knowledge because to be able to apply, assimilate and use external knowledge from a *knowledge-based view*, SMEs need prior related knowledge and *dynamic capabilities* (cf. Cohen & Levinthal 1990, p. 128 f.). The competences gained from *dynamic capabilities* enable firms to develop routines and processes that allow analyzing, processing, interpreting, and understanding of knowledge acquired from external sources, to develop and refine those routines that facilitate combining existing knowledge with acquired and assimilated knowledge for future use and to refine, extend, and leverage existing routines, competencies and technologies or to create new ones by incorporating acquired and transformed knowledge into firms' own operations.

Hurley & Hult (1998) have noted that learning enhances the capacity to understand new ideas and the creativity and ability to notice novel opportunities (cf. Hurley & Hult 1998, p. 46). Learning is inevitable for firms' ability to recognize apparently incongruous knowledge and then to combine it to arrive at a new schema (cf. Zahra & George 2002, p. 190). During the development of new ideas, it is important to keep in mind that *organizational learning* theory suggests that "the generation of new organizational knowledge is maximized in domains close to the domain of existing knowledge, in conditions under which there are few existing organizational routines to unlearn and organizational assimilation and subsequent retrieval of the knowledge occurs in an intense and repetitive fashion" (Autio et al. 2000, p. 911).

According to *Minbaeva et al.* (2003), "the key element in knowledge transfer is not the underlying (original) knowledge, but rather the extent to which the receiver acquires potentially useful knowledge and utilizes this knowledge in [its] own operations" (Minbaeva et al. 2003, p. 587). Those authors have stated that employees' ability and motivation constitute a firm's absorptive capacity and have concluded "that managers can improve the absorptive capacity of their organizations by applying specific HRM practices oriented towards employees' ability (training and performance appraisal) and employees' motivation (performance-based compensation and internal communication)" (Minbaeva et al. 2003, p. 597) in the scope of *managerial cognition.*

In the theory of organization-environment *coevolution* companies, employees and their environments are viewed as the interdependent outcome of managerial actions, institutional influences, and extra-institutional changes (cf. Lewin et al. 1999, p. 535). Absorptive capacity influences the relationships among managerial actions, institutional influences and extra-institutional changes and the organizations, their populations and their environments during *coevolution*, along with how they successfully transform knowledge by supporting learning receptivity.

The open *innovation* model is an approach to address the challenges of innovation, but it does not resolve the basic problem of innovation because the fundamental problem of innovation is not to find more new ideas: the problem of innovation is the establishment of processes and routines that are suitable for the transformation of new knowledge (cf. Denning 2005, p. 8). Transformation is accomplished when knowledge is added or deleted or simply interpreted differently, and learning enables people to complement existing and new knowledge and to merge, synthesize and transform acquired knowledge into higher-value products and services. Therefore, a positive relationship between learning receptivity and the transformation capability can already be assumed. Furthermore, creativity and the will to take action and solve problems is fostered by learning receptivity if there is an established organizational culture that shows employees how to address ideas and errors, which in turn has an impact on their perception of the consequences of errors. This perception affects employees' willingness to report errors and is important for their ability both to identify and analyze problems and to develop new ideas (cf. Sollberger 2006, p. 126).

Based on the literature's theoretical statements on absorptive capacity (especially on transformation) and organizational culture (especially on learning receptivity), and based on organizational learning innovation, dynamic capabilities, the knowledge-based view, managerial cognition and coevolution as theories informing absorptive capacity, it is assumed that learning receptivity is positively related to the transformation capability.

> *Hypothesis H17: Learning receptivity is positively related to the transformation capability.*

3.3.3.6 The Relationship Between Care and the Transformation Capability

In the context of the dimension of care and following the idea of *managerial cognition* that an organization's form — increasingly characterized by flexibility and adaptability — is a management tool in the alignment of an organization and its environment, *Dijksterhuis et al.* (1999) have proposed a *coevolutionary perspective* in which the "contextual variation of macrolevel management logics is proposed as a key mediator in the coevolution of organization and environment" (Dijksterhuis et al. 1999, p. 569). *Dijksterhuis et al.* (1999) have developed a coevolutionary model that shows how contextual applications of management logics can be a source of variation in organizational forms (cf. Dijksterhuis et al. 1999, p. 569). From a *knowledge-based view*, these new organizational forms are inevitable for transformation, where a firm has the capability to develop and refine those routines that facilitate complementing existing knowledge with acquired and assimilated knowledge for future use through developing and increasing that firm's knowledge base (cf. Flatten et al. 2011a, p. 100, cf. Zahra & George 2002, p. 190).

Based on the literature's theoretical statements on absorptive capacity (especially on transformation) and organizational culture (especially on care), and based on the knowledge-based view, managerial cognition and coevolution as theories informing absorptive capacity, it is assumed that care is positively related to the transformation capability.

Hypothesis H18: Care is positively related to the transformation capability.

3.3.3.7 Summary of the Relationship Between Organizational Culture and the Transformation Capability

To summarize, the six dimensions of organizational culture are positively related to the exploitation capability.

3 A Model of External Knowledge Absorption 121

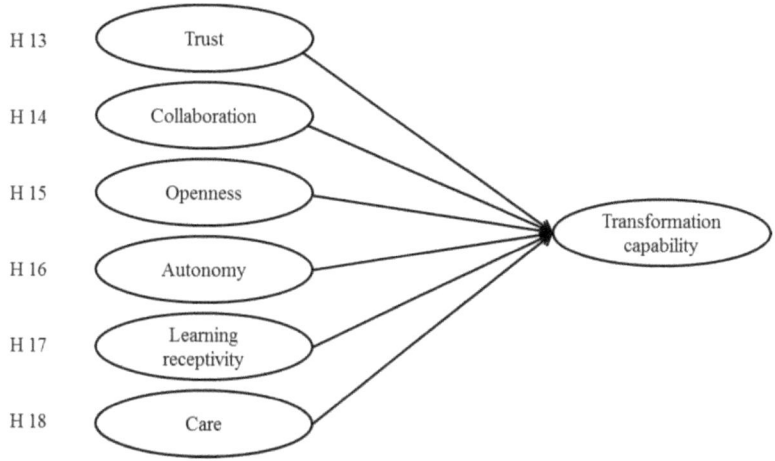

Figure 16 Research Model of the Transformation Capability[81]

Organizational Theories Informing the Transformation Capability

With regard to the organizational theories informing the transformation capability, Table 10 shows that

- **Organizational learning** informs one dimension of organizational culture (learning receptivity),
- **Innovation** informs three dimensions of organizational culture (collaboration, openness and learning receptivity),
- **Dynamic capabilities** inform two dimensions of organizational culture (collaboration and learning receptivity),
- The **knowledge-based view** informs six dimensions of organizational culture (trust, collaboration, openness, autonomy, learning receptivity and care),
- **Managerial cognition** informs four dimensions of organizational culture (trust, collaboration, learning receptivity and care) and

---

[81] Author's own figure.

- **Coevolution** informs two dimension of organizational culture (learning receptivity and care).

Table 10 Theories Informing the Transformation Capability[82]

	Organizational Learning	Innovation	Dynamic Capabilities	Knowledge-Based View	Managerial Cognition	Coevolution
Trust				X	X	
Collaboration		X	X	X	X	
Openness		X		X		
Autonomy				X		
Learning Receptivity	X	X	X	X	X	X
Care				X	X	X
Sum	1	3	2	6	4	2

The transformation capability is most-informed by the knowledge-based view and managerial cognition and is least-informed by organizational learning. The amount of all of the other theories' influence is nearly the same.

With regard to the most important theories informing absorptive capacity, the knowledge-based view is very important for the transformation capability because one detailed aspect of transformation capability is developing and refining routines and processes[83]. The knowledge-based view shows that the processes and routines through which a company transforms knowledge are

---

[82] Author's own table.
[83] The detailed aspects of the transformation capability are explained in section 3.2.1.3.

3 A Model of External Knowledge Absorption   123

fundamental to its ability to create and sustain competitive advantage (cf. Grant 1996b, p. 384).

Managerial cognition is also very important for the transformation capability because one detailed aspect of transformation capability involves combining existing knowledge with acquired and assimilated knowledge for future use. From a managerial recognition point of view, managers can reduce the complexity of absorbing new knowledge. This facilitates the development and refinement of routines and processes, e.g., of those routines that facilitate combining existing knowledge with acquired and assimilated knowledge for future use. The dominant logic can have consequences for absorptive capacity because firms that apply classical management logic will limit their capacity to absorb knowledge.

3.3.4 The Relationship Between Organizational Culture and the Exploitation Capability

3.3.4.1 The Relationship Between Trust and the Exploitation Capability

Trust is very important for the exploitation of external knowledge because increased knowledge transfer caused by trust results in knowledge creation (cf. Choi 2002, p. 51). Trust increases knowledge transfer and knowledge creation: the more knowledge firms acquire through increased knowledge transfer, the more firms can exploit that knowledge by either refining, extending and leveraging existing routines, competencies and technologies or creating new ones by incorporating transformed knowledge into their own operations (cf. Flatten et al. 2011, p. 100, cf. Zahra & George 2002, p. 190). Furthermore, trust guides action by suggesting routines and processes that are most viable under the assumption that the recipient will not exploit the sender's vulnerability, or vice versa (cf. McEvily et al. 2003, p. 93). Therefore, trust encourages a climate that is conducive to the exploitation of knowledge because it alleviates the fear of risk and uncertainty (cf. Choi 2002, p. 51). Furthermore, trust is critical for exploiting knowledge because withholding knowledge due to a lack of trust can be especially harmful to knowledge articulation, internalization and reflection, which is particularly important for knowledge exploitation (cf. Choi 2002, p. 51). Without trust, the exploitation of external knowledge will fail regardless of how thoroughly it is supported by technology and rhetoric (cf. Davenport & Prusak 1998, p. 34), because in an environment that lacks trust, knowledge either will not be created or it will be created in a restrictive manner (cf. Choi 2002, p. 51). Therefore, without trust, the firm's knowledge resource, which according to the *knowledge-based view* is the main determinant of competitive advantage, cannot be built.

Based on the literature's theoretical statements on absorptive capacity (especially on exploitation) and organizational culture (especially on trust), and based on the knowledge-based view as theory informing absorptive capacity, it is assumed that trust is positively related to the exploitation capability.

> *Hypothesis H19: Trust is positively related to the exploitation capability.*

3.3.4.2 The Relationship Between Collaboration and the Exploitation Capability

From a *knowledge-based view*, the creation of new knowledge during collaboration to develop and increase a firm's knowledge base is an essential part of exploitation operations (cf. Flatten et al. 2011a, p. 100, cf. Zahra & George 2002, p. 190). To create new knowledge as an outcome of exploitation routines, it is essential for employees to collaborate because the exchange of knowledge among different employees is a prerequisite for knowledge creation (cf. Lee & Choi 2003, p. 190). Therefore, knowledge creation is fostered by collaborative interactions (cf. Choi 2002, p. 50).

3 A Model of External Knowledge Absorption

Miles et al. (1998) have supported the necessity of collaboration within knowledge exploitation because it is a collaborative process, and knowledge-based approaches cannot succeed without effective collaboration (cf. Miles et al. 1998, p. 286). *Malhotra, Gosain & El Sawy* (2005) have also emphasized the importance of collaboration for the exploitation of knowledge. Their findings have suggested that the exchange of information between partners can lead to new knowledge creation (cf. Malhotra et al. 2005, p. 176).

From a *dynamic capabilities* point of view, collaboration is very important for the exploitation of external knowledge because it depends on links across individual capabilities (cf. Cohen & Levinthal 1990, p. 133). *Nonaka* (2007) has highlighted the role of teams within collaborations: "Teams play a central role in the knowledge-creating company because they provide a shared context where individuals can interact with each other and engage in the constant dialogue on which effective reflection depends. Team members create new points of view through dialogue and discussion. They pool their information and examine it from various angles. Eventually, they integrate their diverse individual perspectives into a new collective perspective" (Nonaka 2007, p. 171). This new, collective perspective can encourage people either to refine, extend and leverage existing routines, competences and technologies or to create new ones by incorporating acquired and transformed knowledge into their own operations, and therefore into their firms' transformation capabilities (cf. Flatten et al. 2011, p. 100, cf. Zahra & George 2002, p. 190).

Probst et al. (2010) has described the team as the most frequent origin of collective knowledge (cf. Probst et al. 2010, p. 126). With regard to *dynamic capabilities*, an organization's absorptive capacity is not resident in any single employee but rather, depends on collaboration among employees because links across individual capabilities are crucial (cf. Cohen & Levinthal 1990, p. 133).

Based on the literature's theoretical statements on absorptive capacity (especially on exploitation) and organizational culture (especially on collaboration), and based on dynamic capabilities and the knowledge-based view as theories informing absorptive capacity, it is assumed that collaboration is positively related to the exploitation capability.

> Hypothesis H20: Collaboration is positively related to the exploitation capability.

3.3.4.3 The Relationship Between Openness and the Exploitation Capability

Furthermore, the exploitation of knowledge requires sharing relevant knowledge among members of a firm to promote mutual understanding (cf. Zahra & George 2002, p. 194)[84]. For sharing knowledge it is necessary to have openness that enables members to orient themselves at a company and to access external fields of knowledge because from a *knowledge-based view* this fosters synergies, cooperation, etc., and the company's internal and external resources are more efficiently used for *innovation* (cf. Probst et al. 2010, p. 65). Furthermore, from a coevolutionary perspective, openness is a critical factor in industrial competitiveness and enables firms both to exploit external knowledge and to predict more accurately the nature of future technological advances (cf. Cohen & Levinthal 1994, p. 227). Therefore, from a *managerial cognition* perspective, dominant logic is an emergent property of organizations as complex adaptive systems (cf. Bettis & Prahalad 1995, p. 14).

Based on the literature's theoretical statements on absorptive capacity (especially on exploitation) and organizational culture (especially on openness) and based on innovation, the knowledge-based view, managerial cognition and coevolution as theories informing absorptive capacity, it is assumed that openness is positively related to the exploitation capability.

---

[84] *Zahra & George* (2002, S. 194), referencing *Garvin* (1993) and *Spender* (1996).

3 A Model of External Knowledge Absorption 127

> *Hypothesis H21: Openness is positively related to the exploitation capability.*

3.3.4.4 The Relationship Between Autonomy and the Exploitation Capability

The relationship between autonomy and exploitation must also be investigated because it can be assumed that autonomy is positively related to the exploitation capability: "By allowing people to act autonomously, the organization may increase the possibility of introducing unexpected opportunities" (Nonaka 1994, p. 18)[85]. The exploitation of these opportunities is very important for the exploitation capability because it denotes a firm's capability to create something new. *Zahra & George* (2002) have specified these new things because they clarify that the outcomes of exploitation routines can be new products, processes, knowledge or organizational forms (cf. Flatten et al. 2011a, p. 100[86], cf. Zahra & George 2002, p. 190). Therefore, autonomy can increase the exploitation of knowledge and have a positive effect on the exploitation capability because the possibility of introducing unexpected opportunities has an impact on the outcome of exploitation. *Molina & Lloréns-Montes* (2006) have supported this assumption and have given this explanation following a *knowledge-based view*: "Autonomy will enhance the probability of creating new knowledge though recombination due to better knowledge flows" (Molina & Lloréns-Montes 2006, p. 267 f.).

Nonaka (1994) has made the following findings in an investigation of the influence of autonomy on exploitation as part of the absorption of external knowledge: "Individual autonomy widens the possibility that individuals will motivate themselves to form new knowledge. Self-motivation based on deep

---

[85] *Nonaka* (1994, p. 18), referencing the work of *Cohen, March & Olsen* (1972) on 'A garbage can model of organizational choice'.

[86] *Flatten et al.* (2011a, p. 190), referencing *Del Carmen Haro-Domínguez, Arias-Aranda, Javier Lloréns-Montes & Ruíz Moreno* (2007).

emotions, for example, in the poet's creation of new expressions, serves as a driving force for the creation of metaphors. A sense of purpose and autonomy becomes important as an organizational context. Purpose serves as the basis of conceptualization. Autonomy gives individuals freedom to absorb knowledge" (Nonaka 1994, p. 18). Furthermore, autonomy allows managers to lay the foundations of the underlying principles of *co-evolving*, self-renewing organizations, e.g., to manage internal rates of change, optimize self-organization and balance concurrent exploitation (cf. Volberda & Lewin 2003, p. 2111).

Based on the literature's theoretical statements on absorptive capacity (especially on exploitation) and organizational culture (especially on autonomy), and based on the knowledge-based view and coevolution as theories informing absorptive capacity, it is assumed that autonomy is positively related to the exploitation capability.

> *Hypothesis H22: Autonomy is positively related to the exploitation capability.*

3.3.4.5 The Relationship Between Learning Receptivity and the Exploitation Capability

Learning receptivity is inevitable for the exploitation of external knowledge because within exploitation the primary emphasis is on the routines, competencies and technologies that allow firms to exploit knowledge: the presence of such routines, competencies and technologies allows firms to specifically exploit knowledge over extended periods of time by providing structural, systemic and procedural mechanisms (cf. Flatten et al. 2011b, p. 139, cf. Zahra & George 2002, p. 190). Furthermore, exploitation denotes a firm's capacity to create something new by being learning-receptive for harvesting and incorporating transformed knowledge into its own operations to create *dynamic capabilities*. Employees must be willing to learn new routines, competencies and technologies: they must be willing to change. This change can be problematic, as *Bettis*

& *Prahalad* (1995) have shown in discussing the questions of why many institutions find change to be so difficult and why many institutions see change in the environment but are unable to act. They have summed up as follows: "Often the focus in trying to answer such questions has been on the surface architecture of the organization strategy, structure, and systems instead of underlying structures and foundations, such as the dominant logic, that support the visible features" (Bettis & Prahalad 1995, p. 7). Therefore, from a *managerial cognition* perspective, the concept of dominant logic can be useful in developing a much more thorough understanding of underlying structures and foundations. From a *coevolutionary perspective*, adaption of structures and foundations is very important because firms increase their absorptive capacity not only by increasing the level of prior related knowledge but also by deliberately changing their organizational forms (cf. Huygens et al. 2001, p. 552).

As a result of exploitation, employees can create new goods, systems, processes, knowledge and organizational forms (cf. Zahra & George 2002, p. 190). From a *knowledge-based view*, *Grant* (1996b) has shown that "the processes through which firms integrate specialized knowledge are fundamental to their ability to create and sustain competitive advantage" (Grant 1996b, p. 384).

With regard to the knowledge creation that is part of the definition of exploitation, *Choi* (2002) has made the following observation related to *organizational learning*: "The mere presence of traditional training and development activities may not be sufficient. Organizations that are serious about knowledge creation need to support a continuous learning environment [...]. Learning should happen at all levels of the organization structure. Individuals must be encouraged to ask questions, to challenge and to learn. This continuous learning opens up the possibility of achieving scale in knowledge creation" (Choi 2002, p. 52). This increasing of knowledge creation through learning fosters the exploitation of external knowledge. Therefore, learning receptivity is inevitable for the exploitation capability and is positively related to it.

Based on the literature's theoretical statements on absorptive capacity (especially on exploitation) and organizational culture (especially on learning receptivity), and based on organizational learning, dynamic capabilities, the knowledge-based view, managerial cognition and coevolution as theories informing absorptive capacity, it is assumed that learning receptivity is positively related to the exploitation capability.

> *Hypothesis H23: Learning receptivity is positively related to the exploitation capability.*

3.3.4.6 The Relationship Between Care and the Exploitation Capability

Care is visible in the courage that people exhibit towards each other (cf. von Krogh 1998, p. 138). This courage is very important to *managerial cognition* in that it empowers employees to give an opinion, propose an idea or give feedback (cf. Sollberger 2006, p. 127). From a *knowledge-based view*, this is in turn very important for either refining, extending and leveraging existing routines, competencies and technologies or creating new ones by incorporating acquired and transformed knowledge into a firm's own operations to generate competitive advantage because renewals must be borne by entire teams and therefore, their opinions should be considered.

Based on the literature's theoretical statements on absorptive capacity (especially on exploitation) and organizational culture (especially on care), and based on the knowledge-based view and managerial cognition as theories informing absorptive capacity, it is assumed that trust is positively related to the exploitation capability.

> *Hypothesis H24: Care is positively related to the exploitation capability.*

3.3.4.7 Summary of the Relationship Between Organizational Culture and the Exploitation Capability

To summarize, the six dimensions of organizational culture are positively related to the exploitation capability.

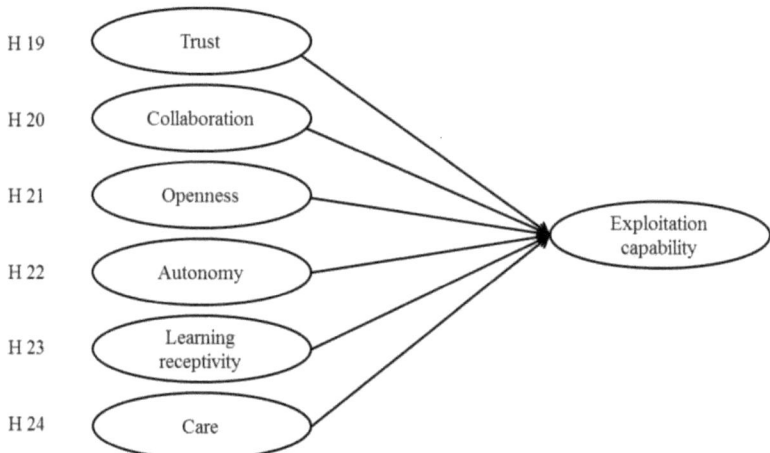

Figure 17 Research Model of the Exploitation Capability[87]

Organizational Theories Informing the Exploitation Capability

With regard to the organizational theories informing the exploitation capability, Table 11 shows that

- **Organizational learning** informs one dimension of organizational culture (learning receptivity),
- **Innovation** informs one dimension of organizational culture (openness),

---

[87] Author's own figure.

- **Dynamic capabilities** inform two dimensions of organizational culture (collaboration and learning receptivity),
- The **knowledge-based view** informs six dimensions of organizational culture (trust, collaboration, openness, autonomy, learning receptivity and care),
- **Managerial cognition** informs three dimensions of organizational culture (openness, learning receptivity and care) and
- **Coevolution** informs three dimensions of organizational culture (openness, autonomy and learning receptivity).

Table 11 Theories Informing the Exploitation Capability[88]

	Organizational Learning	Innovation	Dynamic Capabilities	Knowledge-Based View	Managerial Cognition	Coevolution
Trust				X		
Collaboration			X	X		
Openness		X		X	X	X
Autonomy				X		X
Learning Receptivity	X		X	X	X	X
Care				X	X	
Sum	1	1	2	6	3	3

The exploitation capability is primarily informed by the knowledge-based view, followed by managerial cognition and coevolution. Of the remaining theories, dynamic capabilities, organizational learning and innovation inform the exploitation capability only slightly.

---

[88] Author's own table.

With regard to the most important theories informing absorptive capacity, the knowledge-based view is the most important for the exploitation capability because a detailed aspect of the exploitation capability is refining, extending, and leveraging existing routines, competencies and technologies[89]. In doing this, companies create new knowledge, which is their most important resource because it is the main determinant of competitive advantage.

Managerial cognition is also very important for the exploitation capability because a detailed aspect of the exploitation capability is reflecting. Managers, like knowledge workers, are able to reflect and reduce complexity by using mental maps developed through experience in their core business, which sometimes is inappropriately applied to other businesses (cf. Prahalad & Bettis 1986, p. 485).

Coevolution is also very important for the exploitation capability because a detailed aspect of the exploitation capability is creating new routines, competencies and technologies by incorporating transformed knowledge into a firm's own operations. From a coevolution point of view, it is very important for absorptive capacity that firms ensure that strategic and organization adaptations, e.g., of processes and routines during the exploitation of knowledge, coevolve with changes in the environment because therefore, the absorptive capacity coevolves with changes in the environment. This exploitation capability is critical for competitive advantage because it enables firms to predict more accurately the nature of future technological advances. It is very important for companies not to wait passively for clearer signals from the environment, but to actively exploit knowledge from recognized future technological advances that correspond to their needs.

The following sections summarize the findings about the relationship among the dimensions of trust, collaboration, openness, autonomy, learning

---

[89] The detailed aspects of the exploitation capability are explained in section 3.2.1.4.

receptivity and care, along with the capabilities of acquisition, assimilation, transformation and exploitation.

3.4 The Summarized Presentation of the Model

After developing the several relationships among the dimensions of organizational culture and the capabilities of absorptive capacity in the previous sections, Figure 18 compiles the entire developed hypothesis about how the several dimensions of organizational culture affect the capabilities of absorptive capacity. The results show that all of the dimensions of organizational culture seem to be positively related to each of the four capabilities of absorptive capacity.

With respect to the organizational theories that inform absorptive capacity, Table 12 shows that the knowledge-based view informs all four capabilities of absorptive capacity in the same way, namely, each of the six relationships among one capability of absorptive capacity and the six dimensions of organizational culture. In summary, the knowledge-based view informs absorptive capacity 24 times. Furthermore, Table 12 shows that managerial cognition is the theory that is the second-most informative of each of the four capabilities of absorptive capacity: It informs acquisition six times, assimilation three times, transformation four times and exploitation three times. In summary, managerial cognition informs absorptive capacity 16 times. Organizational learning, innovation, dynamic capabilities and coevolution all play nearly the same, smaller role for each of each of the four capabilities of absorptive capacity and inform absorptive capacity in summary to nearly the same extent. The influence on the several capabilities of absorptive capacity is evenly distributed: Organizational learning is informed between one and three times, innovation is informed between one and three times, dynamic capabilities are informed between one and two times and coevolution is informed between zero and three times. The sums of how often each theory informs absorptive capacity are nearly the same: Organizational learning informs absorptive capacity seven times, innovation informs absorptive capacity nine times, dynamic capa-

3 A Model of External Knowledge Absorption

bilities inform absorptive capacity six times and coevolution informs absorptive capacity seven times.

Table 12 Theories Informing the Capabilities of Absorptive Capacity[90]

	Organizational Learning	Innovation	Dynamic Capabilities	Knowledge-Based View	Managerial Cognition	Coevolution
Acquisition Capability	3	3	1	6	6	2
Assimilation Capability	2	2	1	6	3	0
Transformation Capability	1	3	2	6	4	2
Exploitation Capability	1	1	2	6	3	3
Sum	7	9	6	24	16	7

---

[90] Author's own table.

Acquisition Capability

- H1 Trust is *positively* related to the acquisition capability.
- H2 Collaboration is *positively* related to the acquisition capability.
- H3 Openness is *positively* related to the acquisition capability.
- H4 Autonomy is *positively* related to the acquisition capability.
- H5 Learning receptivity is *positively* related to the acquisition capability.
- H6 Care is *positively* related to the acquisition capability.

Assimilation Capability

- H7 Trust is *positively* related to the assimilation capability.
- H8 Collaboration is *positively* related to the assimilation capability.
- H9 Openness is *positively* related to the assimilation capability.
- H10 Autonomy is *positively* related to the assimilation capability.
- H11 Learning receptivity is *positively* related to the assimilation capability.
- H12 Care is *positively* related to the assimilation capability.

Transformation Capability

- H13 Trust is *positively* related to the transformation capability.
- H14 Collaboration is *positively* related to the transformation capability.
- H15 Openness is *positively* related to the transformation capability.
- H16 Autonomy is *positively* related to the transformation capability.
- H17 Learning receptivity is *positively* related to the transformation capability.
- H18 Care is *positively* related to the transformation capability.

Exploitation Capability

- H19 Trust is *positively* related to the exploitation capability.
- H20 Collaboration is *positively* related to the exploitation capability.
- H21 Openness is *positively* related to the exploitation capability.
- H22 Autonomy is *positively* related to the exploitation capability.
- H23 Learning receptivity is *positively* related to the exploitation capability.
- H24 Care is *positively* related to the exploitation capability.

Figure 18 Summary of Hypotheses H_1 through H_{24}[91]

---

[91] Author's own figure.

Because the knowledge-based view and managerial cognition play the most important role for each of the capabilities of absorptive capacity, absorptive capacity on the whole is primarily informed by the knowledge-based view, followed by managerial cognition. The reason why the knowledge-based view and managerial cognition each play this important role for the acquisition, assimilation, transformation and exploitation capabilities are discussed at the end of each chapter with respect to the relationship between the capability and the theories informing absorptive capacity in detail. From a knowledge-based view, it is very important for absorptive capacity that the main determinant of competitive advantage is a firm's ability to absorb knowledge and develop its knowledge base. Knowledge is the most strategically important of a firm's resources. From a managerial recognition view, it is very important for absorptive capacity that managers can reduce the complexity of absorbing new knowledge. This facilitates the development and refinement of routines and processes, e.g., those routines that facilitate combining existing knowledge with acquired and assimilated knowledge for future use.

The remaining theories — innovation, coevolution, organizational learning and dynamic capabilities — inform each capability of absorptive capacity less than the knowledge-based view and managerial cognition, because they inform only partial aspects of the capabilities of absorptive capacity. Nevertheless, in toto they play an important role for explaining the complex concept of absorptive capacity, because they all make contributions to the capabilities of absorptive capacity.

The results of the empirical analysis of the hypotheses regarding the relationship between the dimensions of organizational culture and the capabilities of absorptive capacity developed in this section will be discussed in the following section.

4 An Empirical Analysis of the Research Models

The objective of the empirical analysis is to evaluate the model of SMEs' external knowledge absorption. The overall research design to pursue this objective can be summarized as follows: In the previous sections, the hypotheses were derived from theoretical statements made in the literature on absorptive capacity and organizational culture and expanded to aspects of the theoretical approaches to organizational theory. An empirically testable model of external knowledge absorption in the context of organizational culture was then developed, which represents how the several dimensions of organizational culture are related to the acquisition, assimilation, transformation and exploitation capabilities at the organizational level. The hypotheses about the relationship among the dimensions of organization culture and the capabilities of absorptive capacity that are represented in the model are the basis of the analysis of the research model in section 4. Figure 19 presents the structure for the empirical analysis of the research models.

To carry out the empirical analysis of the research model, section 4.1 defines SMEs as the object of study (**object of study**).

The methodology of data collection for the empirical analysis of the research model is explained in section 4.2. The hypotheses about the relationships among the dimensions of organizational culture and the capabilities of absorptive capacity represented in the model are tested via the quantitative research method of a survey, allowing for the systematic capture of the SMEs' organizational culture of (**methodology of data collection**).

The model of external knowledge absorption of SMEs is a hypothetical construct and is not directly measurable. For the measurement, it is necessary both to operationalize several components of the model in section 4.3 (**operationalization of the variables**) and to implement operationalized variables in a questionnaire in section 4.4 (**questionnaire**). Subsequently, the sample is described in section 4.5 to obtain an overview of the data (**description of the sample**) and analyzed descriptively in section 4.6 to make points related both to the sample's agreement with the capabilities of the absorptive capacity and the determinations of the organizational culture, along with the normal distribution (**descriptive analysis**).

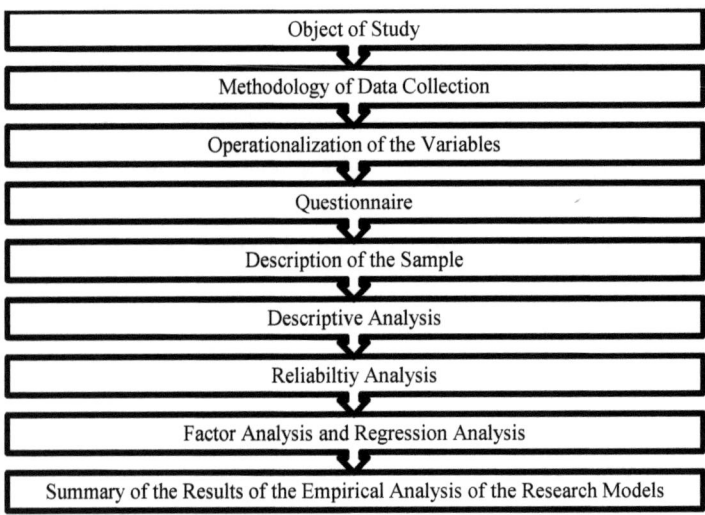

Figure 19 Structure of the Empirical Analysis

For the further empirical analysis, the consistency of the measures must be analyzed by a reliability analysis because there "is virtual consensus among researchers that for a scale to be valid and possess practical utility, it must be reliable" (Peterson 1994, p. 381). Therefore, the sample is analyzed with regard to its reliability (**reliability analysis**). Section 4.8 explains why the sample is analyzed via both a factor analysis (**factor analysis**) and a regression analysis (**regression analysis**).

4 An Empirical Analysis of the Research Models 141

Finally, the results are summarized in section 4.9 (**summary of the results of the empirical analysis of the research models**).

4.1 Object of Study

German SMEs are the objects of study of this thesis. The choice of SMEs as the main context is not arbitrary. SMEs are chosen as the study's objects because they make a significant contribution to Germany's economy. SMEs are extremely important because they provide the foundation for the entire national economy (cf. Institute for SME Research Bonn 2012, p. 12 ff.). Therefore, their economic contribution cannot be overemphasized. Given the strategic importance of the SME sector to growth and job creation, SMEs' success in absorbing external knowledge is critical to the future of the German economy (cf. Lawrence 2010, p. 37).

Various indicators have been used to define SMEs, but research and the literature have also highlighted SMEs' definitional problems (cf. Ndiege et al. 2012, p. 5). Due to the wide diversity of enterprises, there is no single definition of an SME (cf. Culkin & Smith 2000, p. 146). *Ndiege et al.* (2012) have explained in more detail why there is no agreement on a common definition: "Organisations vary in size, location, nature of business, financial performance, maturity and management. These variations make it impossible for various researchers to agree on an objective measure to apply when defining small enterprises. Their definitions vary from country to country with different criteria used to classify them in order to address each country's political and economic purposes" (Ndiege et al. 2012, p. 5).

The *Institute for SME Research Bonn* (2012) has defined independent enterprises with up to 9 employees and less than €1 million in annual turnover as small businesses; moreover, independent enterprises that are not small businesses but have up to 499 employees and an annual turnover of less than €50 million are classified as medium-sized enterprises. Thus, small- and medium-sized enterprises are enterprises with fewer than 500 employees and an annual turnover of less than €50 million. This definition of an SME has been used by the Institute for SME Research Bonn since January 1, 2002. The *European Commission* (2012) and several other organizations also have provided definitions of SMEs.

In this thesis, SMEs are defined in accordance with the definition of the *Institute for SME Research Bonn* (2012). The focus of this thesis is Germany, and using the definition of the *Institute for SME Research Bonn* (2012) allows a comparison of its results with those of other studies in Germany, which primarily use this definition in the context of studies specific to Germany.

4.2   Methodology of Data Collection

The research question of this thesis involves the investigation of how organizational culture is related to the absorption of external knowledge at the organizational level of SMEs, with the purpose of designing a knowledge-oriented organizational culture to ensure the success of SMEs' external knowledge absorption. To achieve the objective of this thesis, quantitative research methods (e.g., Creswell 2009) from the empirical social research have been identified as the most advantageous approach.

Organizational culture and absorptive capacity are considered as organizational constructs that are characterized by collectivism, and individual psychological aspects are excluded from the empirical investigation (cf. Kilian 2005, p. 111). Therefore, the questioning of key informants via questionnaires makes sense as a data-collection method[92]. The reason for the questioning of key informants is that in the context of quantitative, large-scale, empirical investigations, researchers are frequently confronted by a lack of archival data on

---

[92] The questionnaire is described in section 4.4.

4 An Empirical Analysis of the Research Models   143

organization-level constructs of interest and therefore must frequently rely on reports of key informants (cf. Kumar, Stern & Anderson 1993, p. 1633). "Relying on key informant accounts is appropriate when the content of inquiry is such that complete or in-depth information cannot be expected from representative survey respondents" (Kumar et al. 1993, p. 1633 f.). In other words, respondents describe their personal feelings, opinions and behaviors but do not — in contrast to informants — generalize (cf. Kumar et al. 1993, p. 1634)[93]. "The 'use of informants' really means asking the person contacted to act in an informant role. This role involves giving reports about patterns of behavior, after summarizing either observed (actual) or expected (prescribed) organizational relations" (Seidler 1974, p. 817). Therefore, informants are not chosen to be representative of the members of a studied organization in any statistical sense, but because "they are supposedly knowledgeable about the issues being researched and able and willing to communicate about them" (Kumar et al. 1993, p. 1634; cf. Kumar et al. 1993, p. 1634).

The participants in this thesis's survey should be either executives or members of the executive board, because they must be able to provide correct and appropriate responses to statements and questions about knowledge transfers and absorption and about organizational culture.

The survey was distributed through the following channels: forwarding by institutions (e.g., chambers of industry and commerce, business development corporations and marketing clubs) to their members, inclusion in databases (e.g., Key Technologies in Bavaria), publishing as a topic in XING groups (e.g., topics such as SME consulting, SME growth, and SME management), posting to institutions' Facebook pages (e.g., the Association of German Chambers of Industry and Commerce and Bundesverband mittelständische Wirtschaft e.V.) and publishing by institutions (e.g., chambers of industry and commerce and marketing clubs) on their websites or in their newsletters.

---

[93] *Kumar et al.* (1993), referencing *Seidler* (1974).

4.3 Operationalization of the Variables

First, based on the working definition of absorptive capacity as a set of routines and processes that enable firms to acquire, assimilate, transform and exploit knowledge (cf. Zahra & George 2002, p. 186), section 4.3.1 explains and operationalizes the construct of absorptive capacity by reference to its four capabilities of acquisition, assimilation, transformation and exploitation. Secondly, it has been stated that the values of a knowledge-friendly organizational culture are trust, collaboration, openness, autonomy, learning receptivity and care (cf. Sollberger 2006, p. 119). Based on this working definition of organizational culture, the construct of organizational culture is explained and operationalized in section 4.3.2 by reference to its six dimensions of trust, collaboration, openness, autonomy, learning receptivity and care.

4.3.1 Operationalization of Absorptive Capacity

Even though a considerable number of empirical studies have used absorptive capacity, the empirical studies have not developed or validated a multidimensional construct of absorptive capability (cf. Wang & Ahmed 2007, p. 38). A majority of researchers have agreed that absorptive capacity is the extent of prior knowledge in a firm (cf. Lane et al. 2006, p. 844) and have operationalized absorptive capacity using variables considered to be proxies for the extent of prior knowledge, such as R&D intensity (e.g., Meeus, Oerlemans, & Hage 2001, Mowery, Oxley & Silverman 1996, Tsai 2001) and patents (e.g., Ahuja & Katila 2001, Mowery et al. 1996)[94]. However, the appropriateness and validity of such proxies for absorptive capacity are questionable, because those proxies potentially lead to differing results (cf. Lane et al. 2006)[95]. In a similar vein, *Wang &*

---

[94] These examples are taken from *Lane et al.* (2006, p. 844).
[95] *Lane et al.* (2001) have provided the following example: "For instance, Tsai (2001) found support for the influence of R&D intensity (absorptive capacity) in affecting innovation, in contrast to Mowery et al. (1996) and Meeus et al. (2001), who found

Ahmed (2007) have indicated that a significant number of empirical studies have used R&D intensity as a proxy for absorptive capacity and have questioned the extent to which this measure actually reflects its multidimensional nature (cf. Wang et al. 2007, p. 38).

In this thesis, the four capabilities of absorptive capacity — acquisition, assimilation, transformation and exploitation — are operationalized, based on the operationalization of *Flatten et al.* (2011). *Flatten et al.* (2011) have highlighted that most researchers have measured absorptive capacity with simple R&D proxies such as R&D outputs (e.g., patents) and inputs (e.g., R&D intensity) and have ignored the variety of its dimensions and their implications for different organizational outcomes (cf. Flatten et al. 2011, p. 98 f.). In their study, *Flatten et al.* (2011) have developed and validated a multidimensional measure of absorptive capacity, following the established process of item generation and scale development (e.g., Churchill 1979, DeVellis 2011). They have built on the relevant prior literature and have reviewed every article that was published in ten Management Journals (Academy of Management Journal, Academy of Management Review, Administrative Science Quarterly, Journal of Management, Journal of Management Studies, Management Science, Organization Science, Strategic Management Journal, MIS Quarterly, and European Management Journal) from 1990 to 2007, along with further relevant studies that they found in databases, to identify related research streams that are similar to at least one dimension of absorptive capacity (cf. Flatten et al. 2011, p. 100 f.). *Flatten et al.* (2011) have described the importance of this similarity as follows: "An important preliminary step in the development of an initial item pool was the matching of related research streams to a particular ACAP dimension. Therefore, we compared the characteristics derived from the most prominent definitions of each research area with the characteristics of each ACAP dimen-

---

that it was not a good predictor of interorganizational learning" (Lane et al. 2006, p. 844).

sion in order to assign the research streams to one or more ACAP dimensions" (Flatten et al. 2011, p. 101).

The literature review obtained the following results: "33 of the 269 papers in the related research streams contain items that pertain to at least one dimension of ACAP and that could be useful in developing the ACAP scale. The other 236 studies were either of a theoretical or qualitative nature or did not provide any items that were relevant to any ACAP dimension. From these 33 studies we chose an initial item pool of 52 items for their relevance, uniqueness, and ability to convey to informants 'different shades of meaning' of the ACAP construct" (Flatten et al. 2011, p. 105)[96]. Based on these results, *Flatten et al.* (2011) conducted three pre-tests to assess the quality of the 52 items and reduced the number of items to 36 (cf. Flatten et al. 2011, p. 98). These 36 items were tested in two large, survey-based studies of German companies. The results showed that three items described the capability of acquisition, four items described the capability of assimilation, four items described the capability of transformation and three items described the capability of exploitation. Table 13 summarizes these items of the four capabilities of absorptive capacity.

---

[96] *Flatten et al.* (2011, p. 105), referencing *Churchill* (1979, p. 68).

4 An Empirical Analysis of the Research Models

Table 13 Summary of the Operationalization of Absorptive Capacity[97]

	Item
Acquisition Capability	The search for relevant information concerning our industry is every-day business in our company.
	Our management motivates the employees to use information sources within our industry.
	Our management expects that the employees deal with information beyond our industry.
Assimilation Capability	In our company ideas and concepts are communicated cross-departmental.
	Our management emphasizes cross-departmental support to solve problems.
	In our company there is a quick information flow, e.g., if a business unit obtains important information it communicates this information promptly to all other business units or departments.
	Our management demands periodical cross-departmental meetings to interchange new developments, problems, and achievements.
Transformation Capability	Our employees have the ability to structure and to use collected knowledge.
	Our employees are used to absorb new knowledge as well as to prepare it for further purposes and to make it available.
	Our employees successfully link existing knowledge with new insights.
	Our employees are able to apply new knowledge in their practical work.
Exploitation Capability	Our management supports the development of new products and services[98].
	Our company regularly reconsiders technologies for the production of products and services and adapts them accordant to new knowledge[99].
	Our company has the ability to work more effective by adopting new technologies for the production of products and services[100].

The construct of 'absorptive capacity' in its totality is composed as shown in Figure 20.

---

[97] Author's own table, referencing *Flatten et al.* (2011, p. 110). The items are back-translated for the questionnaire in German. The source version is in English and the target version is in German.
[98] Adapted from *Flatten et al.* (2011, p. 110).
[99] Adapted from *Flatten et al.* (2011, p. 110).
[100] Adapted from *Flatten et al.* (2011, p. 110).

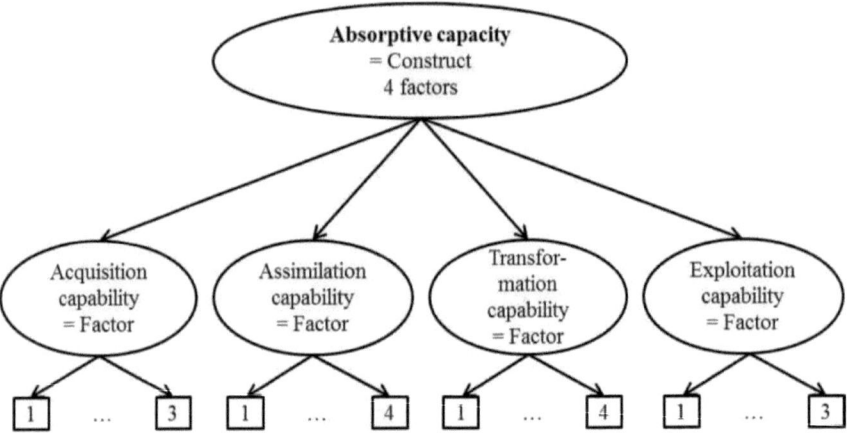

Figure 20 Construct of Absorptive Capacity[101]

In the following section 4.3.2, the six dimensions of organizational culture — trust, collaboration, openness, autonomy, learning receptivity and care — are operationalized.

4.3.2 Operationalization of Organizational Culture

To define a knowledge-friendly organizational culture, *Sollberger* (2006) has worked out the dimensions of organizational culture, which play a central role in the inclusion and integration of holistic knowledge management and therefore, knowledge absorption, with the help of a detailed literature review (cf. Sollberger 2006, p. 115 ff.). The results of *Sollberger*'s (2006) literature review show that trust, collaboration, openness, autonomy, learning receptivity and care are the dimensions of a knowledge-friendly organizational culture, as is shown in section 3.2.2. In this thesis, the six dimensions of organizational culture are

---

[101] Author's own figure.

4 An Empirical Analysis of the Research Models

operationalized based on *Sollberger* (2006)[102]. The reasons for using *Sollberger*'s (2006) dimensions have already been explained in section 2.2.3.

In the context of the factor analysis, *Sollberger* (2006) has eliminated those items that could not be assigned to a factor or where no sufficiently clear assignment was possible (cf. Sollberger 2006, p. 184 ff.). This thesis uses all of *Sollberger*'s (2006) items for operationalization, because the eliminated items could provide appropriate values within this survey.

Three items are used to determine the degree to which organizational culture is characterized by trust. Eight items are used to determine the degree to which organizational culture is characterized by collaboration. Four items are used to determine the degree to which organizational culture is characterized by openness. Two items are used to determine the degree to which organizational culture is characterized by autonomy. Five items are used to determine the degree to which organizational culture is characterized by learning receptivity. Two items are used to determine the degree to which organizational culture is characterized by care. Table 14 and

Table 15 summarize the items of all of the dimensions of organizational culture. These items are back-translated for operationalization in English. The source version is in German and the target version is in English. The items have been adapted according to this thesis's unit of analysis, namely, the organiza-

---

[102] In her own work, *Sollberger* (2006) has investigated knowledge management in the public sector, where knowledge management has a different importance than in the private sector (cf. Sollberger 2006, p. 4). *Sollberger* (2006) has compared the importance of knowledge management in the public sector with the importance of knowledge management in the private sector (cf. Sollberger 2006, p. 4). *Sollberger*'s (2011) research object of study was the Swiss Post. She investigated selected organizational units of that organization. A total of five business units ('ExpressPost', 'PaketPost', 'Paketzentren', 'PostFinance' and 'PostMail') were analyzed, all of which have external client contact but are confronted with competition to a variable extent (cf. Sollberger 2001, p. 165 ff.).

tion. *Sollberger* (2006) has sometimes differentiated between organizations, business units and teams[103].

Table 14 Summary of the Operationalization of Organizational Culture (I)[104]

	Item
Trust	In our company, the superiors lead by example.
	In our daily working environment, the objectives of the company are accepted by all of us.
	In our daily working environment, the skills of the personnel are appreciated as being an important source of competitive advantages.
Collaboration	In our company, we work together efficiently.
	In our daily working environment, we endeavor to find solutions that are beneficial to all persons involved whenever we are of different opinions.
	In our daily working environment, it is even possible to find common ground on how to approach difficult topics and problems.
	In our company, it is easy to coordinate projects that are run by several teams.
	In our company, different teams often work together to achieve joint improvements.
	In our company, we actively support cooperation between different teams (e.g., production and distribution, EDP/IT, finances and personnel etc.).
	In our company, we feel as a part of a big team.
	In the daily work of our company, we put more emphasis on teamwork than on hierarchies. *(adapted)*
Openness	In our company, problems are addressed openly.
	In our company, there are clear objectives, which determine our daily work and lead the way.
	In our company, we have a comprehensible strategy for the future of our company.
	Internal company information about important changes and decisions are communicated in a comprehensible way. *(adapted)*

---

[103] "Business" and "team" are replaced by "company" and "my" is replaced by "our".
[104] Author's own table, referencing *Sollberger* (2006, p. 182).

4 An Empirical Analysis of the Research Models   151

Table 15 Summary of the Operationalization of Organizational Culture (II)[105]

	Item
Autonomy	I know what I am responsible for.
	In our company, problems rarely arise because we have the skills that are required for our jobs.
Learning receptivity	In our daily working environment, information provided by our clients have a direct influence on our decisions.
	In our daily working environment, hints and recommendations provided by our clients often lead to changes.
	In our daily working environment, mistakes are considered to be opportunities to learn and improve.
	In our daily working environment, problems are taken up and processed.
	In our daily working environment, working procedures are reviewed and improved.
Care	In our company, we help each other.
	My immediate superior supports and encourages me.

The construct of 'organizational culture' in its totality is composed as shown in Figure 21.

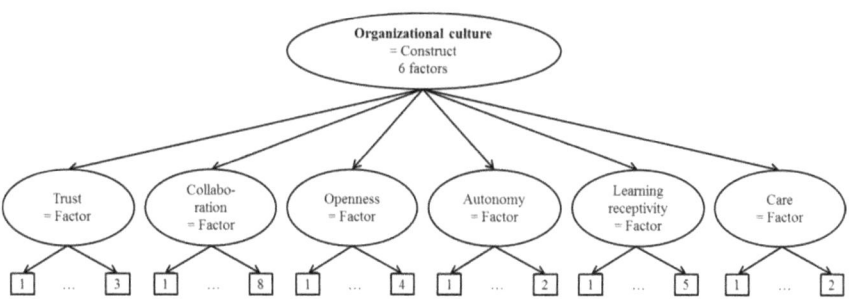

Figure 21 Construct of Organizational Culture[106]

---

[105] Author's own table, referencing *Sollberger* (2006, p. 182).

4.4 Questionnaire

After the capabilities of absorptive capacity and the dimensions of organizational culture were operationalized, the questionnaire could be filled with the items. The questionnaire was divided into three blocks:

1. The **first block** included statements regarding the theoretical construct of absorptive capacity. The statements were arranged in a manner that was analogous to the operationalization of the items of the capabilities of absorptive capacity in section 4.3.1. Evaluation of the items of absorptive capacity was performed using a Likert scale that consisted of a statement followed by a five- or seven-point scale indicating various levels of agreement with the statement (cf. Leroy 2011, p. 208). For the purposes of this thesis, a Likert-type seven-point scale was used to create a more fine-grained scale.
2. The **second block** included statements regarding the theoretical construct of organizational culture. The statements were arranged in a manner that was analogous to the operationalization of the items of the dimensions of the organizational culture in section 4.3.2. As in the first block, evaluation of the items of organizational culture was performed using a Likert-type seven-point scale.
3. The **third block** included questions regarding the studied companies and their socio-demographic characteristics (industrial sector, size, sales, business unit and job position). The several industry branches were extracted from the classification of the branches of economy using the *Federal Office of Statistics* (2008).

---

[106] Author's own figure.

4 An Empirical Analysis of the Research Models 153

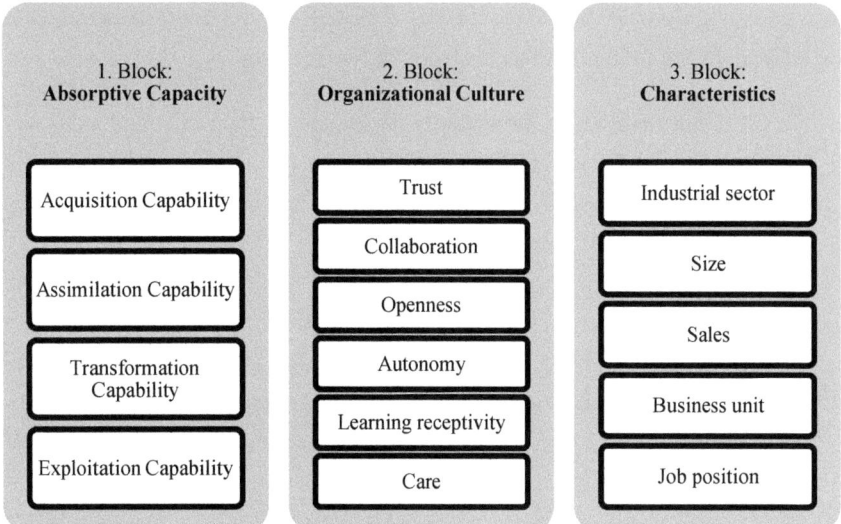

Figure 22 Structure of the Questionnaire[107]

4.5 Description of the Sample

Overall, the empirical study of the relationship between organizational culture and absorptive capacity contained a total sample of 1273 (100.00%) with a response rate of 86.41%, which means that 1,100 persons responded (cf. Table 16)[108]. The difference between the total sample and the persons who responded is explained because there were people who clicked on the survey but then abandoned it. However, only the completed questionnaires were included in the empirical analysis.

The number of completed questionnaires was 526 (a completion rate of 41.32%), with an average process time (mean) of 9 minutes (min) 52 seconds (s)

---

[107] Author's own figure.
[108] Unipark was used as survey software; the field time of the study was 38 days (6/21/2013-7/29/2013).

and an average process time (median) of 6 min 38 s. Fifty-four datasets had to be eliminated from the empirical analysis for two reasons:

1. Certain firms were not actually SMEs due to their size and sales, according to the definition of SME by the *Institute for SME Research Bonn* (2012), because they had more than 500 employees and/or an annual turnover of more than €50 million.
2. Due to their job positions as 'workers' or 'apprentices', certain survey participants were unable to provide correct and appropriate responses to statements and questions about knowledge absorption and organizational culture, because they had not had an adequate overview of absorptive capacity and organizational culture due to a limited scope of work or length of tenure at their companies.

A total of 472 questionnaires (37.08%) were capable of being analyzed.

Table 16 Field Report[109]

Field Report	Absolute Numbers	Percent
Total sample	1273	100.00%
Response rate	1100	86.41%
Completion rate	526	41.32%
Used cases	472	37.08%
Statistical Key Figures		
Average process time (mean)	9 min 52 s	
Average process time (median)	6 min 38 s	

With regard to the analyzed data, the following two tables present the sample's characteristics related to the distribution of industry affiliation and the size of the firms (cf. Table 17).

---

[109] Author's own table.

4 An Empirical Analysis of the Research Models 155

Table 17 Distribution by Industrial Sector[110]

Industrial Sector	Frequency	Percent
Mining and extraction of stones and soils	3	.6
Manufacturing industry	82	17.4
Energy supply; water supply; disposal of wastewater and refuse; removal of environmental pollution	7	1.5
Construction industry	12	2.5
Trade	41	8.7
Transport and storage	5	1.1
Information and communication	107	22.7
Provision of financial and insurance services	2	.4
Housing	2	.4
Provision of freelance, scientific and technical services	58	12.3
Provision of further economic services	52	11
Health and social system	10	2.1
Others	91	19.3
Total	472	100

The distribution of industry affiliation is heterogeneous. With a share of approximately one-fifth (22.7%) of the sample, 'information and communication' is the most represented industry branch, followed by 'others', which also had a one-fifth share (19.3%). Participants who chose the 'others' category were required to state the branch of the industry in which they worked; most often they responded either machine and electrical engineering or IT and software. In total, 'information and communication' and 'others' represented approximately two-fifths of the surveyed firms. Companies from the 'manufacturing industry' (17.4%) represented approximately an additional one-fifth of the sample, and firms involved in the 'provision of freelance, scientific and technical services' (12.3%) and the 'provision of further economic services' (11.00%) each represented only one-tenth of the sample. The remaining branches, which comprised 17.3% of the sample, each represented a small share of less than 9.00% of the sample.

---

[110] Author's own table.

With regard to representativity no statement can be made, because the universal distribution of these industries is unknown. The problem of lack of representativeness is relocatable, because it is not the aim of this partial survey to estimate a single parameter of SMS for the total number of SMEs; rather, this survey attempts to verify the validity of the causal relationships in the theory-based model. In this context, the requirement of representativeness in the survey does not have a high value, as it does in, for example, psephology (cf. Verba, Nie & Kim 1980, p.158 ff.). It is not absolutely required to have a sample that is representative of all SMEs.

The companies surveyed were characterized not only by a generous distribution of industry affiliation, but also by differences in firm size (cf. Table 18). Thirty-five point four percent had 1-9 employees and 40.00% had 10-49 employees; only 21.8% had 50-199 employees and 2.8% had 200-499 employees. Therefore, approximately 75% of the sample consisted of firms that had fewer than 50 employees.

Table 18 Firm Size[111]

Firm Size	Frequency	Percent	Cumulative Percent
1-9	167	35.4	35.4
10-49	189	40	75.4
50-199	103	21.8	97.2
200-499	13	2.8	100
Total	472	100	

The sales of the studied firms were less heterogenic in the first three categories (cf. Table 19). Twenty-three point nine percent had sales of less than €500,000, 23.3% had sales of between €500,000 and €1 million and 36.4% had sales of between €2 million and €9 million. Therefore, 83.7% of the studied firms had

---

[111] Author's own table.

4 An Empirical Analysis of the Research Models 157

sales of less than €10 million. The sales of the remaining firms are distributed as follows: 11.7% had sales of €10 million to €24 million and 4.7% had sales of €25 million to €49 million.

Table 19 Firm Sales[112]

Firm Sales	Frequency	Percent	Cumulative Percent
Less than €500,000	113	23.9	23.9
Between €500,000 and €1 million	110	23.3	47.2
Between €2 million and €9 million	172	36.4	83.7
Between €10 million and €24 million	55	11.7	95.3
Between €25 million and €49 million	22	4.7	100
Total	472	100	

Table 20 represents the respondents' business units. Most of the respondents — 32.8% — said that they work in business units other than those listed in the questionnaire. Among those others, approximately 69% said they worked in management.

With respect to the business units listed in the questionnaire, the units 'others' (32.8%) and 'marketing and sales' (25.8%) were especially dominant, followed by 'EDP and IT' (9.3%) and 'finances and controlling' (9.1%). 'R&D' represented 7.2% and 'production' 5.5% and the remaining branches, which represented 10.2%, had only a small share of the sample with less than 5.00% each.

---

[112] Author's own table.

Table 20 Respondents' Business Units[113]

Respondents' Business Units	Frequency	Percent
Purchase	7	1.5
EDP and IT	44	9.3
Finances and controlling	43	9.1
R&D	34	7.2
Customer service	21	4.4
Logistics	6	1.3
Marketing and sales	122	25.8
Human resources	14	3
Production	26	5.5
Others	155	32.8
Total	472	100

At 88.6% of the participating companies, the questionnaire was filled in by people with a management function: 58.7% described themselves as 'management' and an additional 29.9% were either 'managerial employees' (12.9%), 'team/project supervisors' (8.7%) or 'heads of department' (8.3%; cf. Table 21). In addition to executives and members of the executive board, 11.5% of the respondents held job positions as 'commercial employees' (4.7%), 'technical employees' (1.3%) or 'other' (5.5%).

Due to SMEs' size and structure, it can be assumed that people holding any of these job positions could answer the questions, because they obviously saw themselves as key informants while answering the questions, which were drafted for executives or members of an executive board. Therefore, it can be assumed that the respondents were able to correctly answer the survey's questions about knowledge absorption and organizational culture.

Table 21 Respondents' Job Positions[114]

---

[113] Author's own table.

4 An Empirical Analysis of the Research Models

Respondents' Job Positions	Frequency	Percent
Head of department	39	8.3
Management	277	58.7
Commercial employee	22	4.7
Managerial employee	61	12.9
Team/Project supervisor	41	8.7
Technical employee	6	1.3
Other	26	5.5
Total	472	100

There might be a relationship between a respondent's business unit and his job position, because it can be assumed that in the R&D unit, for example, there are more technical employees than there are in human resources units. To investigate the relationship between a respondent's business unit and his job position, Table 22 and Table 23 calculates a cross-tabulation[115]. The results of the cross-tabulation show the following results with regard to the relationship between a respondent's business unit and his job position.

---

[114] Author's own table.
[115] Business unit (BU) is the independent variable and job position is the dependent variable. Column presents are used.

Table 22 Job Position * Business Unit Cross Tabulation (I)[116]

			Business Unit									
			Purchase	EDP & IT	Finances & Controlling	R&D	Costumer Services	Logistics	Marketing And Sales	Human Resources	Production	Others	Total
Job Position	Head Of Department	Count	2	5	4	6	4	0	9	2	2	5	39
		% with BU	28.6	11.4	9.3	17.6	19	0	7.4	14.3	7.7	3.2	8.3
	Management	Count	4	26	27	13	4	4	60	8	14	117	277
		% with BU	57.1	59.1	62.8	38.2	19	66.7	49.2	57.1	53.8	75.5	58.7
	Commercial Employee	Count	0	1	2	0	3	1	11	0	2	2	22
		% with BU	0	2.3	4.7	0	14.3	16.7	9	0	7.7	1.3	4.7
	Managerial Employee	Count	1	3	8	6	3	1	23	2	3	11	61
		% with BU	14.3	6.8	18.6	17.6	14.3	16.7	18.9	14.3	11.5	7.1	12.9

[116] Own table.

4 An Empirical Analysis of the Research Models 161

Table 23 Job Position * Business Unit Cross Tabulation (II)[117]

			Business Unit									
			Purchase	EDP & IT	Finances & Controlling	R&D	Costumer Services	Logistics	Marketing And Sales	Human Resources	Production	Others	Total
Job Position	Supervisor	Count	0	7	0	5	4	0	15	2	4	4	41
		% with BU	0	15.9	0	14.7	19	0	12.3	14.3	15.4	2.6	8.7
	Technical Employee	Count	0	0	0	2	1	0	2	0	1	0	6
		% with BU	0	0	0	5.9	4.8	0	1.6	0	3.8	0	1.3
	Other	Count	0	2	2	2	2	0	2	0	0	16	26
		% with BU	0	4.5	4.7	5.9	9.5	0	1.6	0	0	10.3	5.5
Total		Count	7	44	43	34	21	6	122	14	26	155	472
		% with BU	100	100	100	100	100	100	100	100	100	100	100

---

[117] Own table.

Head of departments: There were more heads of departments in purchasing (28.6% within BU) than there were heads of department in customer services (19.00% within BU) and R&D (17.6% within BU), which were the two business units with the second largest number of persons who were department heads. The number of department heads in the other business units varied between 14.3% and 7.4%: human resources (14.3% within BU), EDP and IT (11.4% within BU), finances and controlling (9.3% within BU), production (7.7% within BU) and marketing and sales (7.4% within BU). Only in the business unit 'others' (3.2% within BU) and in the business unit logistics (.00% within BU) were fewer than 5.00% department heads.

Management: The 'others' (75.5% within BU) category contained the largest number of respondents who worked in management. Because 69.00% of the respondents that used 'others' said that they worked in management, it can be said that respondents with management positions management were primarily employed as managers. In the business units of logistics (66.7% within BU), finances and controlling (62.8% within BU), EDP and IT (59.1% within BU), human resources (57.1% within BU), purchase (57.1% within BU) and production (53.8% within BU), approximately three-fifths of the respondents in each group held management positions. Even in the other business units, the number of managers was still high: marketing and sales (49.2% within BU), R&D (38.2%) and customer service (19.00% within BU).

Commercial employee: Logistics (16.7% within BU) and customer services (14.3% within BU) were the only business units in which more than 10.00% of the respondents held positions as commercial employees. The proportion of respondents holding commercial-employee positions in the other business units varied between 9.00% (marketing and sales 9.00%, production 7.7%, finances and controlling 4.7%, EDP and IT 2.3% and others 1.3% within BU) and. 00% in the business units of purchasing (.00% within BU), R&D (.00% within BU) and human resources (.00% within BU).

4 An Empirical Analysis of the Research Models   163

Managerial employee: The proportion of respondents holding positions as managerial employees was between 10.00% and 20.00% in nearly every business unit: marketing and sales (18.9% within BU), finances and controlling (18.6% within BU), R&D (17.6% within BU), logistics (16.7% within BU), purchasing (14.3% within BU), customer services (14.3% within BU), human resources (14.3% within BU) and production (11.5%). Only in the business unit 'other' (7.1% within BU) and in the business unit of EDP and IT (6.8% within BU) was the proportion smaller than 10.00%.

Team/Project supervisors: The proportion of respondents holding positions as team/project supervisors was between 10.00% and 20.00% in half of the business units: EDP and IT (15.9% in BU), R&D (14.7% in BU), customer services (19.00% in BU), marketing and sales (12.3% in BU), human resources (14.3% in BU) and production (15.4% in BU). In the other business units, the proportion was smaller than 5.00%: purchasing (.00% in BU), finances and controlling (.00% in BU), logistics (.00% in BU) and 'other' (2.6% in BU).

Technical employees: In R&D (5.9% within BU), customer service (4.8% within BU), marketing and sales (1.6% within BU) and production (3.8% within BU), few respondents worked as technical employees. In the other business units, nobody worked as technical employee.

'Other': 'Other' (10.3% within BU) and customer services (9.5% within BU) were the business units with the largest share of respondents who worked in positions other than the listed ones. In R&D (5.9%), finances and controlling (4.7%) and EDP and IT (4.5%) the share was approximately 5.00% and in the other business units the share was close to .00%: marketing and sales (1.6%), purchase (0.00%), logistics (.00%), human resources (.00%) and production (.00%).

To test the independence of both the business unit in which respondents worked and their job positions, a Chi-Square test was used. A Chi-Square test can apply to any test statistic with a Chi-Square distribution, but it generally refers to Pearson's Chi-Square test (cf. Field 2013, p. 871). The Pearson Chi-Square has a value of 111.154[118] and the significance value test level is .05 (cf. Backhaus, Erichson, Plinke & Weiber 2011, p. 313). If the asymptotic significance (2-sided) of the Pearson Chi-Square statistic is less than .05, then there is a relationship between the business unit in which a respondent worked and his job position. Because the Chi-Square significance value is smaller than .0005, which is less than the significance value test level of .05, there is a relationship between the business unit in which a respondent worked and his job position.

To reduce the problem of empty and nearly empty cells, several business units and job positions are summarized. With regard to business units, purchase, logistics and production are summarized in the new category of 'purchase, logistics and production'. With respect to job positions, commercial employee, managerial employee and technical employee are summarized as the new category of 'commercial, managerial and technical employees'. The Pearson Chi-Square has a value of 85.635[119], and because the Chi-Square significance value is .0005, which is less than the significance value test level of .05, there is a relationship between the business unit in which a respondent worked and his job position[120].

---

[118] "50 cells (71.4%) have expected count less than 5. The minimum expected count is .08." (SPSS output).
[119] "20 cells (50.0%) have expected count less than 5. The minimum expected count is .77." (SPSS output).
[120] The result of the Chi-Square of the dataset with the summarized business units and job positions (20 cells (50.0%) have expected count less than 5) is better than the Chi- Square of the original data set (50 cells (71.4%) have expected count less than 5).

4.6 Descriptive Analysis

First, the capabilities of absorptive capacity and the dimensions of organizational culture are individually considered with regard to the mean[121] and the standard deviation[122]. For absorptive capacity as well as organizational culture, the minimum was 1 ('I totally agree') and the maximum was 7 ('I totally disagree'), which corresponds to the extreme values of the Likert-type 7-point scale (1-7).

With regard to absorptive capacity, the mean of the several variables of the acquisition, assimilation, transformation and exploitation capabilities (cf. Table 24) is primarily in the range of 2, which means that there is a high level of agreement with the several capabilities of absorptive capacity. "Our management emphasizes cross-departmental support to solve problems.", an item of the assimilation capability, has a mean of 1.82 even below 2. The margin of the means is 1.15 and again points out a general agreement with the four capabilities of the absorptive capacity, given the means do not differ greatly.

Measured using a Likert-type 7-point scale, standard deviations between 1.159 and 1.67 can still be counted as small-to-medium standard deviations. In particular, the standard deviation of the transformation capability can be described as small, because three (1.159, 1.238 and 1.276) of the four variables of the transformation capability count to the four smallest of all of the variables of all of the factors. As opposed to this result, the standard deviation of the

---

[121] The mean is a statistical model of the center of a distribution of scores and therefore measures the central tendency of the participants of the survey (cf. Field 2013, p. 22). It hypothetically estimates a typical score (cf. Field 2013, p. 22).

[122] The standard deviation is "an estimate of the average variability (spread) of a set of data measured in the same units of measurement as the original data" (Field 2013, p. 884). "A small standard deviation (relative to the value of the mean itself) indicates that the data points are close to the mean. A large standard deviation (relative to the mean) indicates that the data points are distant from the mean" (Field 2013, p. 27).

acquisition capability is the highest, because all three of its variable variables count to the four highest values (1.44, 1.517 and 1.521).

Table 24 Descriptive Statistics of the Absorptive Capacity (I)[123]

Factor	Variables	N	Mean	Std. deviation
Acquisition Capability	The search for relevant information concerning our industry is every-day business in our company.	471	2.34	1.440
	Our management motivates the employees to use information sources within our industry.	469	2.76	1.521
	Our management expects that the employees deal with information beyond our industry.	470	2.97	1.517
Assimilation Capability	In our company ideas and concepts are communicated cross-departmental.	472	2.47	1.408
	Our management emphasizes cross-departmental support to solve problems.	466	1.82	1.196
	In our company there is a quick information flow, e.g., if a business unit obtains important information it communicates this information promptly to all other business units or departments.	470	2.34	1.334
	Our management demands periodical cross-departmental meetings to interchange new developments, problems, and achievements.	469	2.77	1.670
Transformtion Capability	Our employees have the ability to structure and to use collected knowledge.	470	2.73	1.276
	Our employees are used to absorb new knowledge as well as to prepare it for further purposes and to make it available.	470	2.86	1.391
	Our employees successfully link existing knowledge with new insights.	469	2.64	1.238
	Our employees are able to apply new knowledge in their practical work.	470	2.34	1.159

---

[123] Author's own table; 'std. is an abbreviation of 'standard'.

4 An Empirical Analysis of the Research Models

Table 25 Descriptive Statistics of the Absorptive Capacity (II)[124]

Factor	Variables	N	Mean	Std. deviation
Exploitation Capability	Our management supports the development of new products and services.	469	2.20	1.325
	Our company regularly reconsiders technologies for the production of products and services and adapts them accordant to new knowledge.	469	2.64	1.384
	Our company has the ability to work more effective by adopting new technologies for the production of products and services.	469	2.61	1.328

The distributions of the responses are left-steep and skewed to the right. This suggests a bias toward the positive side. One explanation for this response set is the open and positive understanding of organizational culture emphasized at the beginning of a particular organizational culture and that the respondents strongly identified themselves with their companies. Nevertheless, the bias is harmless because as follows, only the deviations of the measured mean values are of interest, and they are still sufficiently larger than a point on the Likert-type 7-point scale.

With regard to organizational culture, the mean of the several variables of trust, collaboration, openness, autonomy, learning receptivity and care is mostly in the range of 2, which means that a high level of agreement exists. Only "I know what I am responsible for.", an item of autonomy, stands out with a mean of 1.62 and therefore, below 2 (cf. Table 26). Thus, the margin of the means is 1.31, which is slightly higher than the margin of the means of absorptive capacity.

The standard deviation of the several variables of trust, collaboration, openness, autonomy, learning receptivity and care ranges between 1.094 and

---

[124] Author's own table; 'std. is an abbreviation of 'standard'.

1.546. The range of these two extreme values is .452, which is similar to the range of the two extreme values of the absorptive capacity. Trust has the largest number of variables with a value less than or equal to 1.191. Collaboration and openness have the largest number of variables with a value greater than or equal to 1.339.

Table 26 Descriptive Statistics of Organizational Culture (I) [125]

Factor	Variables	N	Mean	Std. Deviation
Trust	In our company, the superiors lead by example.	471	2.20	1.191
	In our daily working environment, the objectives of the company are accepted by all of us.	471	2.36	1.171
	In our daily working environment, the skills of the personnel are appreciated as being an important source of competitive advantages.	471	2.09	1.153
Collaboration	In our company, we work together efficiently.	470	2.61	1.203
	In our daily working environment, we endeavor to find solutions that are beneficial to all persons involved whenever we are of different opinions.	468	2.33	1.207
	In our daily working environment, it is even possible to find common ground on how to approach difficult topics and problems.	468	2.38	1.120
	In our company, it is easy to coordinate projects that are run by several teams.	465	2.82	1.354
	In our company, different teams often work together to achieve joint improvements.	468	2.93	1.546
	In our company, we actively support cooperation between different teams (e.g., production and distribution, EDP/IT, finances and personnel etc.).	468	2.76	1.507
	In our company, we feel as a part of a big team.	466	2.37	1.351
	In the daily work of our company, we put more emphasis on teamwork than on hierarchies.	464	2.23	1.378

Table 27 Descriptive Statistics of Organizational Culture (II) [126]

---

[125] Author's own table.

4 An Empirical Analysis of the Research Models

Factor	Variables	N	Mean	Std. Deviation
Openness	In our company, problems are addressed openly.	464	2.59	1.407
	In our company, there are clear objectives, which determine our daily work and lead the way.	467	2.61	1.339
	In our company, we have a comprehensible strategy for the future of our company.	467	2.76	1.417
	Internal company information about important changes and decisions are communicated in a comprehensible way.	463	2.66	1.399
Autonomy	I know what I am responsible for.	468	1.62	1.094
	In our company, problems rarely arise because we have the skills that are required for our jobs.	465	2.83	1.295
Learning Receptivity	In our daily working environment, information provided by our clients have a direct influence on our decisions.	467	2.13	1.181
	In our daily working environment, hints and recommendations provided by our clients often lead to changes.	467	2.39	1.236
	In our daily working environment, mistakes are considered to be opportunities to learn and improve.	467	2.30	1.270
	In our daily working environment, problems are taken up and processed.	466	2.35	1.149
	In our daily working environment, working procedures are reviewed and improved.	465	2.59	1.248
Care	In our company, we help each other.	469	2.15	1.250
	My immediate superior supports and encourages me.	449	2.29	1.526

The descriptive analysis shows a high level of agreement with the capabilities of absorptive capacity as well as with the determinations of organizational culture.

---

[126] Author's own table.

Another way of looking at the data is to see whether the distribution of scores deviates from a normal distribution. The normal distribution is a probability distribution of a random variable that is perfectly symmetrical and has a kurtosis of 0 (cf. Field 2013, p. 880). A normal distribution can be tested by the Kolmogorov-Smirnov (K-S) test and the Shapiro-Wilk test. This thesis uses the Shapiro-Wilk test, following the arguments of Field (2013): "The K-S test can be used to see if a distribution of scores significantly differs from a normal distribution. If the K-S test is significant […] [($p<.05$)] then the scores are significantly different from a normal distribution. Otherwise, scores are approximately normally distributed. The Shapiro-Wilk test does much the same thing, but it has more power to detect differences from normality (so this test might be significant when the K-S test is not)" (Field 2013, p. 188; own formatting). The results of the Shapiro-Wilk test show that the single variables are not normally distributed because they are left-steep and skewed to the right, because $p<.001$ holds true for the whole analyses of absorptive capacity and organizational culture.

Taking a look at the single histograms of the items, it can be seen that many of them show nearly a normal distribution. Figure 23 is one example of how the histogram of one item of autonomy is presented.

4 An Empirical Analysis of the Research Models 171

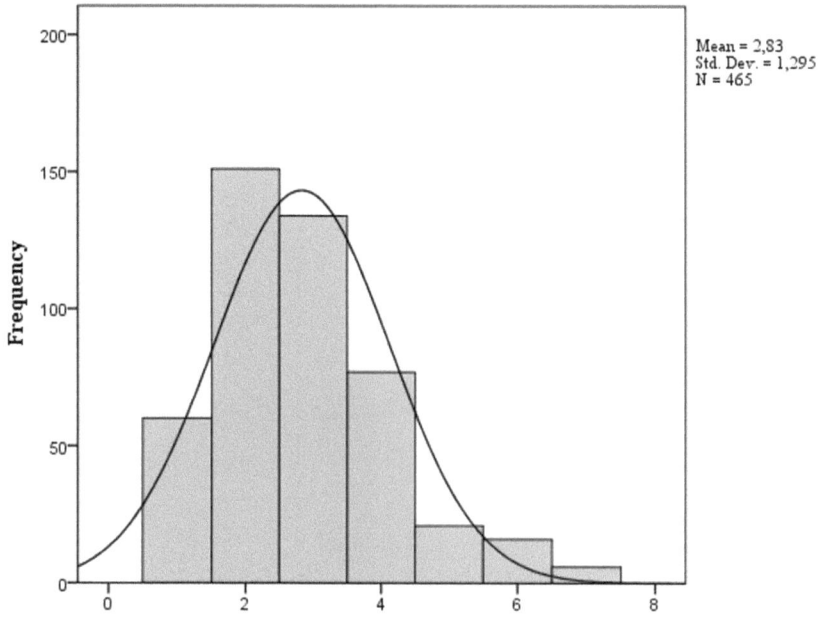

In our company, problems rarely arise because we have the skills that are required for our job.

Figure 23 Histogram: 'In Our Company, Problems Rarely Arise, Because We Have The Skills That Are Required For Our Jobs' (Autonomy)[127]

Figure 24 provides one example of a presentation of the histogram of one item of learning receptivity.

---

[127] Author's own figure.

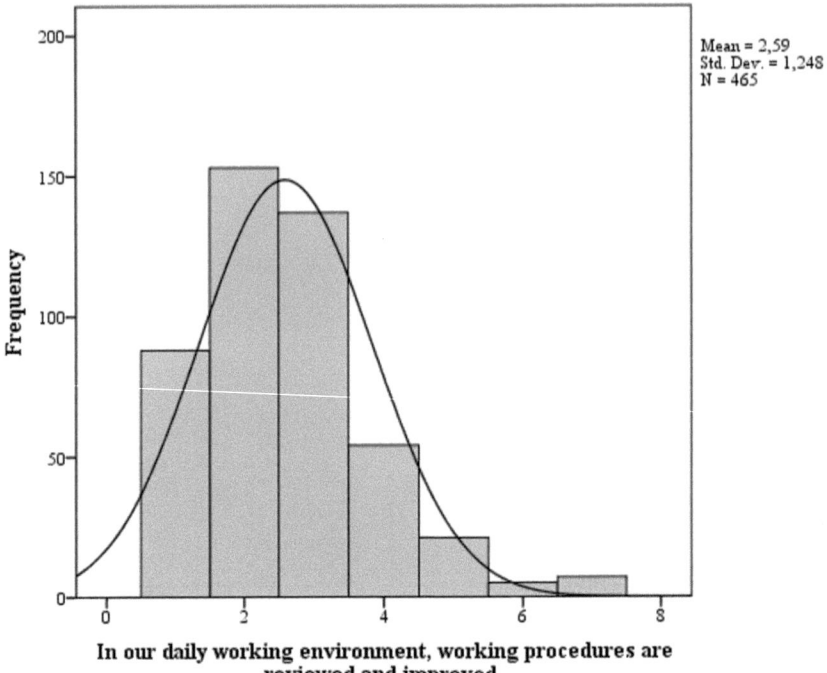

Figure 24 Histogram: 'In Our Daily Working Environment, Working Procedures Are Reviewed And Improved' (Learning Receptivity)[128]

4.7 Reliability Analysis

Reliability is defined as the "similarity of results provided by independent but comparable measures of the same object, trait, or construct" (Churchill 1992, p. 178) and accordingly is a "degree to which measures are free from error and therefore yield consistent results" (Peter 1979, p. 6). A statement about the reliability of the measurement instrument can be made using Cronbach's Alpha (cf.

---

[128] Author's own figure.

4 An Empirical Analysis of the Research Models

Cronbach 1951). Cronbach's Alpha is a reliability coefficient used to test internal consistency reliability, which is an index of reliability for multiple item measures (cf. McKnight, McKnight, Sidani & Figueredo 2007, p. 22). It is regarded as the most recommended measure for calculating the reliability of multi-item scales (cf. Peter 1979, p. 7 ff.). The range of the values of Cronbach's Alpha is between 0 and 1 and values close to 1 express a high degree of reliability (Andrew, Pedersen & McEvoy 2011, p. 202). In general, a high value is desirable to ensure the highest possible quality of the internal consistency of a factor's indicators. A minimum value for this quality measure is controversial in the literature, as shown by the following table with selected recommended reliability levels.

Table 28 Selected Recommended Reliability Levels[129]

Author	Situation	Recommended Level
Davis (1964)	Prediction for individual	Above .75
	Prediction for group of 25-50	.5
	Prediction for group over 50	Below .5
Kaplan & Saccuzzo (1982)	Basic research	.7-.8
	Applied Research	.95
Murphy & Davidshofer (1988)	Unacceptable level	Below .6
	Low level	.7
	Moderate to high level	.8-.9
	High level	.9
Nunnally (1967)	Preliminary research	.5-.6
	Basic research	.8
	Applied research	.9-.95
Nunnally (1978)	Preliminary research	.7
	Basic research	.8
	Applied research	.9-.95

---

[129] Author's own table, referencing *Peterson* (1994, p. 382).

The research of *Davis* (1964), *Kaplan & Saccuzzo* (1982), *Murphy & Davidshofer* (1988), *Nunnally* (1967) and *Nunnally* (1978) has differed with regard to categorizing the situations and the corresponding recommended level and even if they use the same categorizing they differ as to the recommended level. For example, for 'applied research,' *Kaplan & Saccuzzo* (1982) have recommended .95 and *Nunnally* (1967) and *Nunnally* (1978) have recommended .9-.95.

It is often recommended in books and journal articles that a value of .7 to .8 is an acceptable value for the Cronbach's Alpha (cf. Field 2013, p. 709). This value is also considered as relevant for this thesis. If the Cronbach's Alpha of a factor does not reach the required minimum level or optimize the Cronbach's Alpha, the item-total-correlation can be used. Based on the calculation of the strength of the correlation of the indicator variables and the sum of all indicator variables of a construct respectively, with the correlation of an indicator of all the factor's other indicators, a high item to total-correlation indicates a high degree of convergent validity (cf. Nunnally 1978, p. 92 ff.). To increase the Cronbach's Alpha, the indicators with the lowest item- to total-correlations are gradually eliminated (cf. Churchill 1979, p. 69 ff.).

With regard to the absorptive capacity, all of the factors besides the acquisition capability have a Cronbach's Alpha higher than .800 and the Cronbach's Alpha of the transformation capability is higher than .900. The value of the acquisition capability is with .732, which is still above the minimum value of .700.

The 'Cronbach's if Item Alpha Deleted' column shows that removing an item would not improve overall reliability, because no values are greater than the overall reliability indicates and therefore, removing them would not improve the overall reliability. Therefore, all items remain and show with Cronbach's Alpha-coefficients of .732 (acquisition capability), .800 (assimilation capability), .920 (transformation capability) and .832 (exploitation capability) a high reliability of the construct of absorptive capacity.

Table 29 Reliability Statistics of the Absorptive Capacity[130]

Factor	Cronbach's Alpha
Acquisition Capability	.732
Assimilation Capability	.800
Transformation Capability	.920
Exploitation Capability	.832

Table 30 runs a reliability analysis for all of the subscales of the organizational culture. All factors besides autonomy and care have a Cronbach's Alpha higher than .800, and the Cronbach's Alpha of collaboration is higher than .900. The value of care is .707, which is still above the minimum value of .700. Only autonomy, with .625, has a Cronbach's Alpha that is smaller than .700. One reason that autonomy has a Cronbach's Alpha that is smaller than .700 could be that the value of Cronbach's Alpha depends on the number of items on the scale: As the number of items on the scale increases, Cronbach's Alpha will also increase (cf. Field 2013, p. 709). Autonomy has only two items, so they might be the reason for the lower Cronbach's Alpha. Although it is said that values lower than the minimum value indicate an unreliable scale, the .625 value of autonomy is not substantially lower than .7 and still corresponds to the recommended level of *Murphy & Davidshofer* (1988) and *Nunnally* (1967), so it is acceptable for this thesis. Furthermore, lower values are also justified in the case of novel and unexplored topics, which is the case here, as Sollberger's (2006) items are not established in science.

The 'Cronbach's if Item Alpha Deleted' column shows that removing an item would not improve overall reliability. All items remain and show with Cronbach's Alpha-coefficients of .817 (trust), .909 (collaboration), .891 (open-

---

[130] Author's own table.

ness), .879 (learning receptivity) and .707 (care), which show a high reliability of the construct of organizational culture. The Cronbach's Alpha-coefficients of autonomy with .625 is acceptable for this thesis.

Table 30 Reliability Statistics of the Organizational Culture[131]

Factor	Cronbach's Alpha
Trust	.817
Collaboration	.909
Openness	.891
Autonomy	.625
Learning receptivity	.879
Care	.707

To summarize, no values of the measures of absorptive capacity and organizational culture were greater than the overall reliability. Therefore, removing items would not improve the overall reliability of the scales for absorptive capacity and organizational culture. All of the used scales are valid and possess practical utility.

4.8 Factor Analysis and Regression Analysis

A factor analysis and a regression analysis enable the measurement of complex constructs as along with the analysis of complex dependence structures, as has been highlighted by *Spieth* (2009). A factor analysis is a method for the operationalization of latent constructs and a regression analysis is a method for the investigation of the strength of the relationship between one dependent and one

---

[131] Author's own table.

4 An Empirical Analysis of the Research Models

or more independent variables (cf. Spieth 2009, p. 237)[132]. In detail, factor analysis is "a multivariate technique for identifying whether the correlations between a set of observed variables stem from their relationship to one or more latent variables in the data, each of which takes the form of a linear model" (Field 2013, p. 875). "Regression analysis is a statistical tool for evaluating the relationship of one or more independent variables X_1, X_2, ..., X_k to a single, continuous dependent variable Y" (Kleinbaum, Kupper, Nizam & Muller 2008, p. 36). A factor analysis is necessary for this empirical analysis of the research models to identify the linear components of the set of variables and a regression analysis is appropriate because it allows the following applications (cf. Kleinbaum et al. 2008, p. 36):

- To characterize the relationship between a dependent variable (capability of absorptive capacity) and an independent variables (dimension of organizational culture) by determining the extent, direction and strength of the association.
- To describe a capability of absorptive capacity as a function of the dimensions of organizational culture.
- To determine which of the several dimensions of organizational culture are important and which are not important for describing or predicting a capability of absorptive capacity.
- To compare the several derived regression relationships between the dimensions of organizational culture and the capabilities of absorptive capacity.

Although there are more complex methods, such as structural equation models[133] (e.g., Weiber & Mühlhaus 2010), regression analysis is the "most popular

---

[132] *Spieth* (2006, p. 237), referencing *Bagozzi, Fornell & Larcker* (1982) and *Bagozzi* (1979).
[133] "Structural equation modeling represents the hybrid of two separate statistical traditions. The first tradition is factor analysis developed in the disciplines of psychology and psychometrics. The second tradition is simultaneous equation modeling devel-

and the most fundamental technique in applied statistics" (Berk 2004, p. xi). As it is the objective of this thesis to clearly identify the central determinants for the explanation of the dependent variable, the following sections present a factor analysis and subsequently, a regression analysis. A structural equation model puts less stress on the tested hypothesis and instead is often used to generate hypotheses about the relationships among the dimensions of organizational culture and the capabilities of absorptive capacity. Therefore, the results of a structural equation model would leave more hypotheses unrejected and instead modify them, which is not the aim of this thesis.

4.8.1 Factor Analysis

There are a variety of extraction methods for factor analyses, e.g. maximum likelihood factor analysis (ML), image factoring, alpha factoring, generalized least squares and unweighted least squares. Two methods are of particular importance: principal axis factoring (PAF) and principal component analysis (PCA). The PAF and the PCA have three common issues (cf. Stier 1999, p. 273; cf. Field 2013, p. 666 ff.):

- the understanding of the structure of a set of variables;
- the construction of a questionnaire to measure an underlying variable; and
- the reduction of a dataset to a more manageable size while retaining as much of the original information as possible

These methods differ with regard to their interpretation of the factors. The underlying question during the interpretation of the factors during a PAF reads as follows: *'How is it possible to refer to the cause that is responsible for the high correlations between the variables?'*; the underlying question during the inter-

---

oped mainly in econometrics, but having an early history in the field of genetics" (Kaplan 2000, p. 1).

4 An Empirical Analysis of the Research Models

pretation of the factors during a PCA reads as follows: *'How is it possible to summarize by a collective term variables that are highly loading on a factor?'* (cf. Backhaus et al. 2011, p. 357)[134].

The decision on whether to use PAF and PCA is thus determined only by objective and contextual considerations. Because this thesis does not assume that it is not the question of the hypothetical explanation size but instead the question of the comprehensive reproduction of the data structure that is of importance, PCA is used in the following analysis to identify the linear components of the set of variables. Using PCA, the maximum amount of total variance in a correlation matrix can be explained by transforming the original variables in linear components (cf. Field 2013, p. 667).

PCA is used more often than any other method of exploratory factor analysis[135]. In a survey of a recent two-year period in PsycINFO that yielded more than 1,700 studies that used some form of exploratory factor analysis, *Costello & Osborne* (2005) have discovered that more than half of the authors listed PCA as the method used for data analysis (cf. Costello & Osborne 2005, p. 1). Therefore, the majority used the Kaiser criterion (cf. Costello & Osborne 2005, p. 1). The Kaiser-Meyer-Olkin (KMO) criterion, which is also called the Measure of Sampling Adequacy (MSA) criterion, indicates to which extent the initial analysis seems to make sense or not and allows both an overall assess-

---

[134] Translated by the author.
[135] The goal of PCA is a comprehensive reproduction of the data structure by as few factors as possible (cf. Jolliffe 2002, p. 151; cf. Field 2013, p. 666). Therefore no distinction between commonalities and individual residual variance is performed. Therefore PCA is frequently treated as an independent extraction technique (cf. Backhaus et al. 2011, p. 357). *Jolliffe* (2002) has discussed the similarities and differences between PCA and factor analysis and summarizes thusly: "Both PCA and factor analysis aim to reduce the dimensionality of a set of data, but the approaches taken to do so are different for the two techniques. Principal component analysis has been extensively used as part for factor analysis, but this involves 'bending the rules' that govern factor analysis and there is much confusion in the literature over the similarities and differences between the techniques" (Jolliffe 2002, p. 151) A detailed discussion can be found in *Jolliffe* (2002, p. 151 ff.).

ment of the correlation matrix and of the individual variables (cf. Backhaus et al. 2011, p. 342 f.). In the literature, the KMO criterion is referred to as the best available method for testing the correlation matrix (cf. Dziuban & Shirkey 1974, p. 360 f., cf. Backhaus et al. 2011, p. 342 f.). The values of the KMO criterion range between 0 and 1 (cf. Field 2013, p. 677 ff.). *Kaiser* (1974) has suggested the following levels for the evaluation of the KMO value (cf. Table 31).

Table 31 Evaluation of Levels of KMO[136]

Value	Evaluation
In the .90s	Marvelous
In the .80s	Meritorious
In the .70s	Middling
In the .60s	Mediocre
In the .50s	Miserable
Below .50	Unacceptable

A value of KMO greater or equal to .8 is desirable (cf. Kaiser 1970, p. 405). A minimum value for this quality measure is as controversial in the literature as is the Cronbach's Alpha. The minimum values vary between .6 (cf. Stewart 1981, p. 58) and lower requirements such as a factor loading of .3 (cf. Churchill 1991, p. 500 ff., cf. Matijevic 2005, p. 190). In this thesis, .5 is used as the elimination criterion for factor loading, *following* Kaiser (1974).

Table 32 shows the results of the factor analysis of the absorptive capacity. The Bartlett's-Test of the absorptive capacity to examine whether the variance-covariance matrix is proportional to the identity matrix was significant for all factor analyses of absorptive capacity (p<.001), which means that the correlation matrix is significantly different from an identity matrix and there-

---

[136] Author's own table, referencing *Kaiser* (1974, p. 35).

fore, the correlations between variables are significantly different from zero (cf. Field 2013, p. 684).

Corresponding to the KMO criterion, only a factor that explains a substantial amount of variance retains. All of the factor loadings of the single indicators of the capabilities of absorptive capacity are higher than or equal to the generally requested limit of .500, and the factor loading of the transformation capability (.847) is even greater than the desirable value of .8.

The factors explain 65.56% of the total variance of the acquisition capability, 63.91% of the total variance of the assimilation capability, 81.03% of the total variance of the transformation capability and 74.92% of the total variance of the exploitation capability, which is satisfying. For absorptive capacity no factor rotations were necessary because the unrotated data provided the best results, and rotations are only necessary if the unrotated data showed no interpretable results.

Table 32 Factor Analysis of the Absorptive Capacity[137]

Factor	Variables of the Factor	KMO	Total Variance
Aquisition Capability	The search for relevant information concerning our industry is every-day business in our company. (.828) Our management motivates the employees to use information sources within our industry. (.885) Our management expects that the employees deal with information beyond our industry. (.706)	.615	65.56%
Assimilation Capability	In our company ideas and concepts are communicated cross-departmental. (.846) Our management emphasizes cross-departmental support to solve problems. (.849) In our company there is a quick information flow, e.g., if a business unit obtains important information it communicates this information promptly to all other business units or departments. (.758) Our management demands periodical cross-departmental meetings to interchange new developments, problems, and achievements. (.740)	.788	63.91%
Transformation Capability	Our employees have the ability to structure and to use collected knowledge. (.890) Our employees are used to absorb new knowledge as well as to prepare it for further purposes and to make it available. (.894) Our employees successfully link existing knowledge with new insights. (.923) Our employees are able to apply new knowledge in their practical work. (.894)	.847	81.03%
Exploitation Capability	Our management supports the development of new products and services. (.799) Our company regularly reconsiders technologies for the production of products and services and adapts them accordant to new knowledge. (.915) Our company has the ability to work more effective by adopting new technologies for the production of products and services. (.879)	.672	74.92%

[137] Author's own table.

4 An Empirical Analysis of the Research Models

Below, the four factors of absorptive capacity are investigated with respect to their correlation. "A coefficient of + 1 indicates that the two variables are perfectively positively correlated, so as one variable increases, the other increases by a proportionate amount. Conversely, a coefficient of - 1 indicates a perfect negative relationship: if one variable increases, the other decreases by a proportionate amount. A coefficient of zero indicates no linear relationship at all and so if one variable changes, the other stay the same" (Field 2013, p. 267). How the values of the correlation coefficients can be interpreted in detail is shown in the following table. If the value of the correlation coefficient is 0 there is no correlation. A value of the correlation coefficient of >= 0-.2 indicates a very weak correlation, .2-.4 a weak correlation, 4-.6 a middle correlation, .6-.8 a strong correlation, .8-<1 a very strong correlation and 1 a perfect correlation.

Table 33 Interpretation of Correlation Coefficients[138]

Value of the correlation coefficient	Possible interpretation
0	No correlation
>= 0-.2	Very weak correlation
.2-.4	Weak correlation
.4-.6	Middle correlation
.6-.8	Strong correlation
.8-<1	Very strong correlation
1	Perfect correlation

The values of the significance in Table 34 show that all factors correlate with each other on the level of $p<.001$, and the values of the Pearson Correlation show that the assimilation capability, the transformation capability and the ex-

---

[138] Author's own table, referencing *Brosius* (1998, p. 503).

ploitation capability have correlations ranging from mediocre to very significant with each other[139].

The exploitation capability and the assimilation capability (.632**) and the exploitation capability and the transformation capability (.610**) have a strong and positive correlation and the transformation capability and the assimilation capability (.564**), the assimilation capability and the acquisition capability and (.464**) and the exploitation capability and the acquisition capability (.404**) have a middle and positive relationship because all of their Pearson Correlation-coefficients are above .400. The transformation capability and the acquisition capability (.395**) still demonstrate a middle relationship, which is positive. These results show that only the correlation of the acquisition capability with the other capabilities of absorptive capacity is lower than .400. All of the other variables correlate as either middle or strongly positive with each other, with Pearson correlation coefficients higher than .400, which was expected, because the four capabilities of absorptive capacity are ordered chronologically. Another reason for the high correlation could be the response set — specifically, that the respondents determined the "right" answer.

---

[139] Because there is more than one null hypothesis and several questions are answered on the basis of the same data set, it is well-known that multiple testing is a problem: "Multiple testing refers to any instance that involves the simultaneous testing of more than one hypothesis. If decisions about the individual hypotheses are based on the unadjusted marginal p-values, then there is typically a large probability that some of the true null hypotheses will be rejected" (Romano, Shaikh & Wolf 2010, p. 1). Therefore, the significance test is used only as test of quantities and is purely descriptive in this thesis.

4 An Empirical Analysis of the Research Models

Table 34 Correlations of the Absorptive Capacity[140]

		Acquisition Capability	Assimilation Capability	Transformation Capability
Assimilation Capability	Pearson Correlation	.464**		
	Sig. (2-tailed)	.000		
	N	460		
Transformation Capability	Pearson Correlation	.395**	.564**	
	Sig. (2-tailed)	.000	.000	
	N	463	459	
Exploitation Capability	Pearson Correlation	.404**	.632**	.610**
	Sig. (2-tailed)	.000	.000	.000
	N	463	458	461

Note: + = not significant (n.s.), * $p < .05$, ** $p < .01$, *** $p < .001$

Analyzing organizational culture, the Bartlett's-Test is also significant for all of the factor analyses of organizational culture ($p<.001$). Table 35 shows the results for the factor analysis of organizational culture. Here, too, no factor rotations are necessary.

All of the factor loadings of the single indicators of the dimensions of organizational culture are higher than or equal to the generally requested limit of .500, and the factor loadings of collaboration (.871), openness (.825) and learning receptivity (.815) are even greater than the desirable value of .8. Only autonomy and care have exactly .5, which is borderline, but acceptable.

Table 35 Factor Analysis of the Organizational Culture (I) [141]

---

[140] Author's own table.

Factor	Variables of the Factor	KMO	Total Variance
Trust	In our company, the superiors lead by example. (.852) In our daily working environment, the objectives of the company are accepted by all of us. (.870) In our daily working environment, the skills of the personnel are appreciated as being an important source of competitive advantages. (.846)	.716	73.27%
Collaboration	In our company, we work together efficiently. (.813) In our daily working environment, we endeavor to find solutions that are beneficial to all persons involved whenever we are of different opinions. (.814) In our daily working environment, it is even possible to find common ground on how to approach difficult topics and problems. (.778) In our company, it is easy to coordinate projects that are run by several teams. (.778) In our company, different teams often work together to achieve joint improvements. (.741) In our company, we actively support cooperation between different teams (e.g., production and distribution, EDP/IT, finances and personnel etc.). (.757) In our company, we feel as a part of a big team. (.800) In the daily work of our company, we put more emphasis on teamwork than on hierarchies. (.804)	.871	61.82%
Openness	In our company, problems are addressed openly. (.803) In our company, there are clear objectives, which determine our daily work and lead the way. (.888) In our company, we have a comprehensible strategy for the future of our company. (.894) Internal company information about important changes and decisions are communicated in a comprehensible way. (.887)	.825	75.49%

[141] Author's own table.

4 An Empirical Analysis of the Research Models

Table 36 Factor Analysis of the Organizational Culture (II) [142]

Factor	Variables of the Factor	KMO	Total Variance
Autonomy	I know what I am responsible for. (.855) In our company, problems rarely arise because we have the skills that are required for our jobs. (.855)	.500	73.12%
Learning Receptivity	In our daily working environment, information provided by our clients have a direct influence on our decisions. (.803) In our daily working environment, hints and recommendations provided by our clients often lead to changes. (.753) In our daily working environment, mistakes are considered to be opportunities to learn and improve. (.868) In our daily working environment, problems are taken up and processed. (.880) In our daily working environment, working procedures are reviewed and improved. (.806)	.815	67.76%
Care	In our company, we help each other. (.882) My immediate superior supports and encourages me. (.882)	.500	77.83%

The factors explain 73.27% of the total variance of trust, 61.82% of the total variance of collaboration, 75.49% of the total variance of openness, 73.12% of the total variance of autonomy, 67.76% of the total variance of learning receptivity and 77.83% of the total variance of care, which is satisfying.

Below, the six factors of organizational culture are investigated with respect to their correlation. The values of significance in Table 37 show that all of the factors correlate with each other on the level of p<.001, and the values of the Pearson Correlation show that trust, collaboration, openness autonomy,

---

[142] Author's own table.

learning receptivity and care correlate with each other, with Pearson Correlation-coefficients between .653 and .764, and therefore, they are very positively related.

Table 37 Correlations of the Organizational Culture[143]

		Trust	Collaboration	Openness	Autonomy	Learning Receptivity
Collaboration	Pearson Correlation	.718**				
	Sig. (2-tailed)	.000				
	N	450				
Openness	Pearson Correlation	.764**	.749**			
	Sig. (2-tailed)	.000	.000			
	N	456	440			
Autonomy	Pearson Correlation	.674**	.653**	.696**		
	Sig. (2-tailed)	.000	.000	.000		
	N	464	446	452		
Learning Receptivity	Pearson Correlation	.761**	.694**	.725**	.711**	
	Sig. (2-tailed)	.000	.000	.000	.000	
	N	454	438	442	451	
Care	Pearson Correlation	.620**	.747**	.644**	.584**	.602**
	Sig. (2-tailed)	.000	.000	.000	.000	.000
	N	447	430	434	442	432

Note: + = not significant (n.s.), * p < .05, ** p < .01, *** p < .001

[143] Author's own table.

4 An Empirical Analysis of the Research Models 189

First, the results show that all of the dimensions of organizational culture have a positive relationship with each other. The most positive correlations appear for variables related with trust. The Pearson Correlation-coefficient of openness and trust is .764** and of openness and collaboration is .749**. Therefore, their correlation is strong. With regard to openness, the Pearson Correlation-coefficients of learning receptivity and openness with .725**, of autonomy and openness with .696** and of care and openness with .644**, are strong. Care and collaboration with a Pearson Correlation-coefficient of .747** and collaboration and trust with a Pearson Correlation-coefficient of .718** show that trust is followed by collaboration with the highest Pearson Correlation-coefficient. With respect to collaboration, the Pearson Correlation-coefficients of learning receptivity and collaboration with .694** and of autonomy and collaboration with .653** are also strong. The left results are mixed: learning receptivity and autonomy (.711**), autonomy and trust (.674**), learning receptivity and trust (.761**), care and trust (.620**), care and learning receptivity (.602**) and care and autonomy (.584**). Altogether, the dimensions of organizational culture correlate from middle to strong, which was expected because the dimensions of trust, collaboration, openness, autonomy, and learning receptivity also influence each other, because together they build the culture of an organization. Another reason for the high correlation could be the response set.

To summarize, except for care and autonomy, which show a middle correlation, all of the dimensions correlate strongly with each other.

The next section, 4.8.2, conducts a regression analysis, because it is necessary to check the relationships between the dimensions of organizational culture and the capabilities of absorptive capacity to prove or reject the hypotheses that are developed in section 3.3. The testing of the hypothesis with a regression analysis exposes the hypothesis to maximum stress.

4.8.2 Regression Analysis

Four regression models are computed, one for each of the capabilities of absorptive capacity with respect to the six dimensions of organizational culture. Within each regression analysis, the coefficient of determination (R^2), the regression coefficient (B), the standard error of the regression coefficient (SE B) and the standardized regression coefficient (β) are analyzed.

R^2: In a regression analysis, R^2 measures 'goodness of fit', which is an index of how well a model fits the empirical data and is based on how well the predicted data correspond to the data that were actually collected (cf. Backhaus et al. 2011, p. 72, cf. Field 2013, p. 875). R^2 is the proportion of variance in one variable explained by a second variable and is calculated by squaring the Pearson's correlation coefficient (cf. Field 2013, p. 872). R^2 can vary between 0 and 1: Zero indicates that the predictors are useless at predicting the outcome variable and 1 indicates that the model perfectly predicts the outcome variable (cf. Field 2013, p. 765).

B: B is an unstandardized regression coefficient. It is an indicator for "the strength of relationship between a given predictor […] of many and an outcome in the units of measurement of the predictor. It is the change in the outcome associated with a unit change in the predictor" (Field 2013, p. 870).

SE B: SE B is defined as "the standard deviation of the *sampling distribution* of a statistic" (Field 2013, p. 884). It indicates whether a statistic from a given sample is an accurate reflection of the population from which the sample came (cf. Field 2013, p. 884).

β: β is the standardized regression coefficient and indicates how strong a given predictor of many is related to an outcome in a standardized form (cf. Field 2013, p. 870). "It is the change in the outcome (in standard deviations) associated with a one standard deviation change in the predictor" (Field 2013, p. 870).

4 An Empirical Analysis of the Research Models   191

In the following four sections, from section 4.8.2.1 to section 4.8.2.4, the relationship between the dimensions of organizational culture and the acquisition, the assimilation, the transformation and the exploitation capabilities is empirically validated.

4.8.2.1   Organizational Culture and the Acquisition Capability

The R^2 of the acquisition capability is .138 and therefore, the predictors do not predict the outcome variable very well. To interpret this result, the coefficients of acquisition are first calculated and the results interpreted. The values of B in Table 38 indicate that the gradient of all of the several dimensions of organizational culture is positive and that the strength of the relationship between a predictor and the outcome variable is between a minimum absolute value of .039 (openness) and a maximum absolute value of .122 (autonomy). The values of SE B in Table 38 indicate that the statistic from the sample may be an accurate reflection of the population from which the sample came, because the values of SE B are not bigger than .087 and therefore are not high.

Hypothesis H_1 through *hypothesis H_6* cannot be confirmed, because none of the constructs of the six dimensions of organizational culture are significant:

- The trust construct (*hypothesis H_1*) exceeds a β of .059 and is not significant (β = .059, n.s.).
- The collaboration construct (*hypothesis H_2*) exceeds a β of .096 and is not significant (ß = .096, n.s.).
- The openness construct (*hypothesis H_3*) exceeds a β of .039 and is not significant (ß = .039, n.s.).
- The autonomy construct (*hypothesis H_4*) exceeds a β of .124 and is not significant (ß = .124, n.s.).
- The learning receptivity construct (*hypothesis H_5*) exceeds a β of .067 and is not significant (ß = .067, n.s.).

- The care construct (*hypothesis H₆*) exceeds a β of .043 and is not significant (ß = .043, n.s.).

Table 38 Coefficients of the Acquisition Capability[144]

	B	SE B	β
Constant	.009	.047	
Trust	.057	.086	.059⁺
Collaboration	.094	.087	.096⁺
Openness	.039	.086	.039⁺
Autonomy	.122	.073	.124⁺
Learning receptivity	.066	.082	.067⁺
Care	.043	.073	.043⁺

Note: dependent variable: acquisition capability; R^2 = .138; + = not significant (n.s.), * $p < .05$, ** $p < .01$, *** $p < .001$

The finding that the dimensions of organizational culture are not critical for the acquisition capability is unexpected and interesting, because several studies attribute success of the acquisition of external knowledge to the several dimensions of organizational culture, and all of the relationships are theoretically supported by the theories informing absorptive capacity — especially the knowledge-based view and managerial cognition[145]. Nevertheless, the regression analysis of acquisition capability already shows that there is no significant relationship.

The acquisition capability is the only interface with the external environment in the process of the absorption of external knowledge, and the results of the regression analysis show that organizational culture has no influence on

---

[144] Author's own table.
[145] This is shown in section 3.3.1, which discusses the relationships between organizational culture and the acquisition capability.

the acquisition capability. One reason for this result might be that firms are so harshly confronted with their need for resources that they must acquire knowledge and take less care of their artifacts, values, and basic underlying assumptions. SMEs suffer from a lack of resources that are essential to generate long-term competitive advantages (cf. Vanhaverbeke 2012, p. 9). To deal with this resource challenge, firms have increasingly focused on external knowledge absorption instead of operating independently (cf. Ding et al. 2009, p. 48). Interactions with suppliers and customers (e.g., von Hippel 1988), interorganizational relationships, including R&D consortia, strategic alliances and joint ventures (e.g., Baum & Ingram 1998, Darr et al. 1995), observations (e.g., Nonaka et al. 1996), patents (e.g., Appleyard 1996), etc. are examples of external knowledge sources. Usage of these sources has increased in recent decades (cf. Ding et al. 2009, p. 47, cf. Powell et al. 1996, p. 116). The knowledge-based view supports this increase in knowledge acquisition and claims that knowledge is a firm's most important resource because it is the main determinant of competitive advantage (cf. Kogut & Zander 1992, p. 384). Another reason for the rejection of the hypothesis might be that there are such strict rules for the acquisition of external knowledge, in the scope of the usage of sources of external knowledge, that it is impossible for organizational culture to play a role. For example, with respect to patents everything is regulated when a firm acquires knowledge.

Table 39 provides an overview of the results of the examination of *hypothesis H_1* through *hypothesis H_6*. The empirical results do not confirm the expected positive relationship between the dimensions of organizational culture and the acquisition capability.

The first analysis focuses on examining the relationship between the dimensions of organizational culture and the acquisition capability. The result is an accidental product of the gathered data, which means that it does not hold for the population. The next analysis, which is contained in section 4.8.2.2, focuses on examining the relationship between the dimensions of organizational culture and the assimilation capability.

Table 39 Overview of Hypotheses H_1 through H_6[146]

	No.	Hypothesis	Decision
	H_1	Trust is *positively* related to the acquisition capability.	Rejected
	H_2	Collaboration is *positively* related to the acquisition capability.	Rejected
Acquisition Capability	H_3	Openness is *positively* related to the acquisition capability.	Rejected
	H_4	Autonomy is *positively* related to the acquisition capability.	Rejected
	H_5	Learning receptivity is *positively* related to the acquisition capability.	Rejected
	H_6	Care is *positively* related to the acquisition capability.	Rejected

4.8.2.2 Organizational Culture and the Assimilation Capability

The R^2 of the assimilation capability is .568 and indicates that the model predicts the outcome variable satisfactorily. The values of B in Table 40 indicate that the gradient of all of the several dimensions of organizational culture are positive and that the strength of the relationship between a predictor and the outcome variable is between a minimum absolute value of .014 (learning receptivity) and a maximum absolute value of .455 (collaboration). Only the gradient of autonomy is negative, and the strength of the relationship between a predictor and the outcome variable has an absolute value of .122.

The values of SE B in Table 40 indicate that the statistics from the sample may be an accurate reflection of the population from which the sample came, because the values of SE B are not bigger than .063 and therefore are not high. The SE B of the assimilation capability is smaller than the SE B of the acquisition capability (.087).

As shown in Table 40, the trust construct exceeds a β of .195 at a significance level of 1% (ß = .195, p < .01). Accordingly, support is found for *hypothesis H_7*. The finding that trust is highly relevant for the assimilation capability supports the suggestion — based on the literature, the knowledge-based

---

[146] Author's own table.

view and managerial cognition — that employees who trust each other can set the stage for overcoming barriers to the assimilation of external knowledge. This finding is also in line with the expectations derived from the literature in section 3.3.2. The results of the regression analysis show that trust is very important with regard to the assimilation of external knowledge. Trust allows people to conserve knowledge, which according to the *knowledge-based view* is a firm's most important resource because it is the main determinant of competitive advantage (cf. Kogut & Zander 1992, p. 384, cf. McEvily et al. 2003, p. 93). Furthermore, the results of the regression analysis stress the necessity for the sender both to trust that the recipient will deal with the knowledge responsibly and to be certain that the recipient will not abuse the acquired knowledge (cf. Kunz 2010, p. 35). Once again, it is noted that trust represents a positive assumption about another person's motives and intentions within a knowledge transfer between colleagues and therefore, it allows people to economize on knowledge processing and safeguarding behaviors, as has also been highlighted by *McEvily et al.* (2003) highlight (cf. McEvily et al. 2003, p. 92 f.). Thus, the results confirm the expectation that when trust is high, the overall knowledge exchange and the likelihood that the resources acquired from other persons is assimilated increase (cf. Abrams et al. 2003, p. 65). The reason for this increase is that from a *managerial cognition* perspective, trust makes decision-making more efficient by simplifying the acquisition and interpretation of knowledge that corresponds to the assimilation of knowledge (cf. McEvily et al. 2003, p. 93).

Hypothesis H_8 is also supported by the empirical data (ß = .456, p < .001). Thus, the results confirm the expectation, based on the literature and on organizational learning, innovation and the knowledge-based view, that collaboration is positively related to the assimilation capability. So far, the findings show that the assimilation focuses not only on individual performance but also on the collaboration. This finding is also in line with the expectations derived from the literature in section 3.3.2. The importance of involving all employees in the assimilation of external knowledge postulates that firms ensure a collaborative environment for assimilating the externally acquired knowledge, because collaboration is essential for the acceptance of knowledge sharing (cf. O'Dell & Grayson 1999, p. 13 f.). Furthermore, collaboration is essential for the development and understanding of knowledge (cf. Sollberger 2006, p. 123) and therefore, this understanding of knowledge is inevitable for the assimilation of knowledge. Specifically, regarding a firm's capability to develop and refine those routines that facilitate combining existing knowledge with acquired and assimilated knowledge for future use, it is important that collaboration offers new ways of thinking with regard to combining knowledge. For example, the teams can be divided into competing groups that develop different approaches to the same problem and then argue over the advantages and disadvantages of their proposals (cf. Nonaka 2007, p. 168). This encourages people to look at a problem from a variety of perspectives and eventually helps to develop a common understanding of the 'best' approach (cf. Nonaka 2007, p. 168). In summary, the results point out (once again) that is very important that a collaborative environment is created for the assimilation capability — particularly because acquired knowledge can embody heuristics that differ significantly from those used by the company, which can result in delayed comprehension of the knowledge during organizational learning (cf. Zahra & George 2002, p. 189)[147]. *Kogut & Zander* (1992) have supported the assumption that the ability to collaborate and comprehend knowledge within an organizational context is a competitive advantage, which is the focus of the *knowledge-based view* (cf. Kogut & Zander 1992, p. 384).

Hypothesis H_9 is also empirically confirmed, although on a lower significance level than the last hypothesis (ß = .192, p < .01). The findings highlight that openness is of relevance for the assimilation capability as an active exchange of knowledge in an open atmosphere enables employees at all levels to understand the routines and processes in the company what is essential for the assimilation capability (cf. Davenport & Prusak 1998, p. 49, cf. Sollberger 2002, p. 124). The finding is also in line with the expectations derived from the litera-

---

[147] *Zahra & George* (2002, p. 189), referencing *Leonard-Barton* (1995).

4 An Empirical Analysis of the Research Models

ture and the knowledge-based view in section 3.3.2: Openness is essential for the assimilation of external knowledge, because openness enables knowledge sharing between different organizational units and hierarchical levels (cf. Sollberger 2002, p. 124).

Table 40 Coefficients of the Assimilation Capability[148]

	B	SE B	β
Constant	.007	.034	
Trust	.191	.062	.195**
Collaboration	.455	.063	.456***
Openness	.192	.062	.192**
Autonomy	-.122	.053	-.122*
Learning receptivity	.014	.059	.014⁺
Care	.067	.052	.068⁺

Note: dependent variable: assimilation capability; $R^2 = .568$; ⁺ = n.s., * $p < .05$, ** $p < .01$, *** $p < .001$

In contrast to the previous hypotheses, *hypothesis H_{10}* cannot be confirmed, as the β of the autonomy construct is negative and not very significant (β = -.122, $p < .05$). The finding that autonomy is negatively related to the assimilation capability is unexpected and interesting, as several studies attribute the success of the assimilation of external knowledge to employee autonomy (e.g., Pemberton & Stonehouse 2002, Molina & Lloréns-Montes 2006), and the knowledge-based view supports this hypothesis. In those studies, it has been argued that autonomy seems to be particularly important when firms are confronted with situations in which they need prior related knowledge both to recognize, share and communicate important information and to make decisions to foster the assimilation of external knowledge. The reason for the negative relationship may be employ-

---

[148] Author's own table.

ees who have been forced to pool specific, complex knowledge gained from their prior training, because it is normal to require employees to bring together knowledge that is scattered throughout the firm (cf. Molina & Lloréns-Montes 2006, p. 267). Employees might feel put-upon. Nevertheless, the ß is very small and autonomy is barely related to assimilation.

Hypothesis H_{11} also is not supported by the empirical data, because the learning receptivity construct is non-significant (ß = .014, n.s.). Thus, the results reject the expectation — based on organizational learning, dynamic capabilities, the knowledge-based view and managerial cognition — that learning receptivity is positively related to the assimilation capability. This finding is unexpected and contrasts with the findings of several studies. For example, the studies by *Hurley & Hult* (1998), *Flatten et al.* (2011a), *Zahra & George* (2002), *Cohen & Levinthal* (1989) have supported a positive relationship and have argued that if employees are receptive to learning they will understand new ideas, be creative, be able to notice novel opportunities and be able to solve problems (cf. Hurley & Hult 1998, p. 46). A reason for the findings of this thesis might be that employees' creativity and implementability are restricted by company regulations and therefore cannot be acted upon. This might create the feeling of deficient learning.

Hypothesis H_{12} also is not empirically confirmed, because the care construct is also non-significant (ß = .068, n.s.). The findings highlight that care is not relevant to the assimilation capability, although it has been argued in the literature and supported by the knowledge-based view and managerial cognition, as shown in section 3.3.2, that care helps people to learn, to increase the awareness of important events and their consequences, and to nurture personal knowledge creation while sharing insights, which are inevitable for understanding knowledge obtained from external sources and for analyzing, interpreting and understand new knowledge during its assimilation. One reason might be that the employees are so independent that they do not conceive care as a dimension of their organizational cultures.

4 An Empirical Analysis of the Research Models

Table 41 gives an overview of the results of the examination of hypotheses H_7 through H_{12}. The empirical results confirm the expected positive relationship among three of the organizational dimensions, namely trust and the assimilation capability (*hypothesis H_7*), collaboration and assimilation capability (*hypothesis H_8*) and openness and the assimilation capability (*hypothesis H_9*). The hypotheses about the positive relationship between openness and the assimilation capability (*hypothesis H_{10}*), autonomy and the assimilation capability (*hypothesis H_{11}*) and care and the assimilation capability (*hypothesis H_{12}*) are rejected.

Table 41 Overview of the Hypotheses H_7 through H_{12}[149]

	No.	Hypothesis	Decision
	H_7	Trust is *positively* related to the assimilation capability.	Accepted
	H_8	Collaboration is *positively* related to the assimilation capability.	Accepted
Assimilation Capability	H_9	Openness is *positively* related to the assimilation capability.	Accepted
	H_{10}	Autonomy is *positively* related to the assimilation capability.	Rejected
	H_{11}	Learning receptivity is *positively* related to the assimilation capability.	Rejected
	H_{12}	Care is *positively* related to the assimilation capability.	Rejected

The second analysis focuses on examining the relationship between the dimensions of organizational culture and the assimilation capability. The next analysis, in section 4.8.2.3, focuses on examining the relationship between the dimensions of organizational culture and the transformation capability.

---

[149] Author's own table.

4.8.2.3 Organizational Culture and the Transformation Capability

The R^2 of the transformation capability is .506 and indicates that the model predicts the outcome variable very satisfactorily. The values of B in Table 42 indicate that the gradient of all four of the dimensions of organizational culture are positive and that the strength of the relationship between a predictor and the outcome variable is between a minimum absolute value of .089 (autonomy) and a maximum absolute value of .435 (collaboration). Again, collaboration has the maximum absolute value, just as with assimilation. The gradients of openness and care are negative and the strength of the relationship between a predictor and the outcome variable has an absolute value of .072 for openness and an absolute value of .074 for care, which lie between the extreme values of the strength of the positive relationships.

The values of SE B in Table 42 indicate that the statistic from the sample may be an accurate reflection of the population from which the sample came, because the values of SE B are not bigger than .066 and therefore are not high.

As shown in Table 42, the trust construct exceeds a β of .260 at a significance level of .1% (ß = .260, p < .01). Accordingly, support is found for *hypothesis* H_{13}. Thus, the results confirm the expectation that trust is positively related to the transformation capability. This is in line with the expectations derived from the literature, the knowledge-based view and managerial cognition in section 3.3.3. Once again, it is noted that it is important to take note of trust, because trust between the sender and the recipient of knowledge influences the ability to understand new knowledge (cf. Lane et al. 2001, p. 1139) and to develop and refine those routines that facilitate combining existing knowledge with acquired and assimilated knowledge for future use (cf. Flatten et al. 2011a, p. 100, cf. Zahra & George 2002, p. 190).

4 An Empirical Analysis of the Research Models

Table 42 Coefficients of the Transformation Capability[150]

	B	SE B	β
Constant	-.013	.036	
Trust	.254	.066	.260***
Collaboration	.435	.066	.437***
Openness	-.072	.066	-.072⁺
Autonomy	.089	.056	.090⁺
Learning receptivity	.130	.063	.130*
Care	-.074	.055	-.074⁺

Note: dependent variable: transformation capability; $R^2 = .506$; ⁺ = n.s., * $p < .05$, ** $p < .01$, *** $p < .001$

Hypothesis H_{14} is also supported by the empirical data (ß = .437, p < .001). The finding that collaboration is highly relevant for the transformation capability supports the suggestion that employees who collaborate with each other can set the stage for overcoming barriers to transforming external knowledge. This finding is also in line with the expectations derived from the literature and innovation, dynamic capabilities, the knowledge-based view and managerial cognition in section 3.3.3. It supports the initial suggestion that collaboration is necessary to foster knowledge utilization in the sense of merging, synthesizing and transforming acquired knowledge into products and services (cf. Spender 1996, p. 46). The importance of collaboration for the transformation of external knowledge increases the need for an attractive environment. *Pemberton & Stonehouse* (2002) have made a relevant statement with regard to enabling action from a *knowledge-based view*: "While the structure and infrastructure must clearly be in place, without the creation of a cultural environment that facilitates social interaction, organizations are likely to be more that knowledge enabled and, at best partially knowledge managed" (Pemberton & Stonehouse 2002, p. 87). To create a cultural environment that facilitates social interaction within a collaboration, it is important to take note of employees' underlying concerns

---

[150] Author's own table.

related to sharing information. *Davenport et al.* (1997) have given the example that employees might feel that their knowledge is critical to maintaining their value and link that knowledge directly to their job security (cf. Davenport et al. 1997, p. 15). Under these circumstances, employees will be reluctant to share knowledge or take action because they will be afraid that their value and therefore, their job security, is inextricably tied to their personal knowledge and expertise (cf. Davenport et al. 1997, p. 15). There must be an absence of knowledge inhibitors in the organizational culture, and firms must ensure that their employees do not fear that knowledge-sharing will cost them their unique characteristics and jobs (cf. Davenport et al. 1997, p. 15). Furthermore, the results of the regression analysis stress, with regard to *innovation*, that it is important to keep in mind that the fundamental problem in innovation is not to find more new ideas but to establish, refine and leverage routines and processes to collaborate so that the organization is open to exploring new ideas and willing to back the most promising one (cf. Denning 2005, p. 8). This problem is very important with regard to the transformation capability.

In contrast to the previous two hypotheses, *hypothesis H_{15} cannot be confirmed*, because the ß of the openness construct is negative and non-significant (ß = -.072, n.s.). Thus, the results reject the expectation that openness is positively related to the transformation capability. This finding is unexpected and interesting, because the hypothesis is supported by innovation and the knowledge-based view and several studies attribute the success of the transformation of external knowledge to employees' openness (e.g., Katz & Tushman 1981, Allen & Cohen 1969, Badaracco 1991) because what parties are trying to learn from each other or create together is often embedded in the practice and culture of a firm and is difficult to communicate, and transformation is only possible through working relationships that are not hampered by constraints (cf. Badaracco 1991, p. 142). Nevertheless, the literature discusses the NIH syndrome very intensively and the results support the relevance of the problem of negative attitudes towards openness. Openness is very important for transformation because employees must both accept the knowledge that has been developed and refine those routines that facilitate combining existing knowledge with acquired and assimilated knowledge for future use. Employees who believe that they possess a monopoly on knowledge in their area of specialization do not seriously consider the possibility that outsiders might produce important new knowledge relevant to the group and therefore, they do not adapt their routines to new knowledge (cf. Katz & Allen 1982, p. 7). As a consequence, they may completely miss the possibility of the transformation of and improvements to their routines.

Hypothesis H_{16} is also not supported by the empirical data because the autonomy construct is non-significant (ß = .090, n.s.). The finding that autonomy is not critical for the transformation capability is unexpected and interesting, because several studies have attributed the success of the transformation capability to employee autonomy. These studies have argued that employee autonomy in scheduling work and determining routines and processes supports the transformation of external knowledge. Employees must feel free to develop and refine their routines because they are reluctant to turn their knowledge and ideas into action if they must fear being sanctioned for errors (cf. Sollberger 2006, p. 125). Furthermore, the hypothesis is supported by the knowledge-based view. One reason for this result could be that employees have so many possibilities at their companies that they do not feel that they need to act autonomously.

Hypothesis H_{17} is empirically confirmed (ß = .130, p < .05). The finding that learning receptivity is highly relevant for the transformation capability supports the suggestion — based on organizational learning innovation, dynamic capabilities, the knowledge-based view, managerial cognition and coevolution — that employees who are receptive to learning can set the stage for overcoming barriers to the transformation of external knowledge. This finding is also in line with the expectations derived from the literature in section 3.3.3. The importance of learning receptivity postulates that employees are able to acquire potentially useful knowledge and utilize this knowledge in their own operations, because "the key element in knowledge transfer is not the underlying (original) knowledge, but rather the extent to which the receiver acquires potentially useful knowledge and utilizes this knowledge in own operations"

(Minbaeva et al. 2003, p. 587). As *Hurley & Hult* (1998) have found, the results of the regression analysis indicate that learning enhances the capacity to understand new ideas and the creativity and the ability to notice novel opportunities (cf. Hurley & Hult 1998, p. 46) and that learning is inevitable for the ability of firms to recognize apparently incongruous knowledge and then combine it to arrive at a new schema (cf. Zahra & George 2002, p. 190). During the development of new ideas, it is important to keep in mind that *organizational learning* theory suggests that the generation of new knowledge is maximized in domains where there is a knowledge base (cf. Autio et al. 2000, p. 911).

Hypothesis H_{18} is not empirically confirmed because the ß of the care construct is negative and non-significant (ß = -.074, n.s.). Thus, the results reject the expectation that care is positively related to the transformation capability, although the literature and the knowledge-based view, managerial cognition and coevolution support the positive relationship, as shown in section 3.3.3. The reason for the negative relationship may be that care both fosters lenient judgment among members during knowledge sharing and is visible in the courage that members exhibit towards each other (cf. von Krogh 1998, p. 138). This courage is very important for employees to give an opinion, propose an idea or give feedback (cf. Sollberger 2006, p. 127). Nevertheless, if employees give their opinions inconsiderately, rashly propose ideas or give feedback that is too harsh, care can have a negative effect.

Table 43 gives an overview of the results of the examination of *hypothesis* H_{13} through *hypothesis* H_{18}. The empirical results confirm the expected positive relationship between three organizational dimensions — namely, trust and the transformation capability (*hypothesis* H_{13}), collaboration and the transformation capability (*hypothesis* H_{14}) and learning receptivity and the transformation capability (*hypothesis* H_{17}). The hypotheses about the positive relationship between openness and the transformation capability (*hypothesis* H_{15}), autonomy and the transformation capability (*hypothesis* H_{16}) and care and the transformation capability (*hypothesis* H_{18}) are rejected.

4 An Empirical Analysis of the Research Models

Table 43 Overview of the Hypotheses H_{13} through H_{18}[151]

	No.	Hypothesis	Decision
Transformation	H_{13}	Trust is *positively* related to the transformation capability.	Accepted
	H_{14}	Collaboration is *positively* related to the transformation capability.	Accepted
	H_{15}	Openness is *positively* related to the transformation capability.	Rejected
	H_{16}	Autonomy is *positively* related to the transformation capability.	Rejected
	H_{17}	Learning receptivity is *positively* related to the transformation capability.	Accepted
	H_{18}	Care is *positively* related to the transformation capability.	Rejected

The third analysis focuses on examining the relationship between the dimensions of organizational culture and the transformation capability. The next analysis, in section 4.8.2.4, focuses on examining the relationship between the dimensions of organizational culture and the exploitation capability.

4.8.2.4 Organizational Culture and the Exploitation Capability

The R^2 of the exploitation capability is .529 and indicates that the model predicts the outcome variable very satisfactorily. The values of B in Table 44 indicate that the gradient of four of the several dimensions of organizational culture is positive and that the strength of the relationship between a predictor and the outcome variable is between a minimum absolute value of .084 (trust) and a maximum absolute value of .316 (collaboration). Again, collaboration has the maximum absolute value as before, with regard to assimilation and transformation. The gradient of autonomy and care are negative and the strength of the relationship between a predictor and the outcome variable has an absolute value of .098 for autonomy and an absolute value of .013 for care. Therefore, the

---

[151] Author's own table.

extreme values of the B are at a minimum absolute value of .013 for care and a maximum absolute value of .316 for collaboration.

The values of SE B in Table 44 indicate that the statistic from the sample may be an accurate reflection of the population from which the sample came, because the values of SE B are not larger than .065 and therefore are not high.

Table 44 Coefficients of the Exploitation Capability[152]

	B	SE B	β
Constant	.004	.035	
Trust	.084	.064	.086⁺
Collaboration	.316	.065	.319***
Openness	.202	.065	.203**
Autonomy	-.098	.055	-.098⁺
Learning receptivity	.292	.062	.291***
Care	-.013	.054	-.013⁺

Note: dependent variable: exploitation capability; $R^2 = .529$; ⁺ = n.s., * $p < .05$, ** $p < .01$, *** $p < .001$

Hypothesis H_{19} is not supported by the empirical data because the trust construct is non-significant, as shown in Table 44 (β = .086, n.s.). The finding that trust is negatively related to the exploitation capability is unexpected and interesting, because several studies attribute the success of the exploitation capability to employee trust and the hypothesis is supported by the knowledge-based view, as shown in section 3.3.4. Trust is very important for the exploitation capability because the more knowledge that is transferred by trust, the more knowledge is created (cf. Choi 2002, p. 51). According to *Choi* (2002), trust increases knowledge transfer and therefore knowledge creation. The more knowledge

---

[152] Author's own table.

firms acquire through an increased knowledge transfer, the more firms are able to exploit that knowledge either by refining, extending and leveraging existing routines, competencies and technologies or by creating new ones by incorporating new knowledge into their own operations (cf. Flatten et al. 2011, p. 100, cf. Zahra & George 2002, p. 190).

Hypothesis H_{20} is supported by the empirical data (ß = .319, p < .001). Thus, the results confirm the expectation, based on dynamic capabilities and the knowledge-based view, that collaboration is positively related to the exploitation capability. Once again, it is noted that to create new knowledge as an outcome of the exploitation routines, collaboration is essential. This finding is also in line with the expectations derived from the literature in section 3.3.4, which describes the exchange of knowledge among different people as a prerequisite for knowledge creation (cf. Lee & Choi 2003, p. 190). From a *knowledge-based view*, the creation of new knowledge during collaboration to develop and increase a firm's knowledge base is an essential part of the exploitation operations (cf. Flatten et al. 2011a, p. 100, cf. Zahra & George 2002, p. 190). To create new knowledge as an outcome of exploitation routines, collaboration among employees is essential, because knowledge exchange of knowledge among employees is a prerequisite for knowledge creation (cf. Lee & Choi 2003, p. 190). Therefore, knowledge creation is fostered by collaborative interactions (cf. Choi 2002, p. 50). The results of the regression analysis and other authors such as *Miles et al.* (1998) support the necessity of collaboration within knowledge exploitation, because exploitation is a collaborative process and knowledge-based approaches cannot succeed without effective collaboration (cf. Miles et al. 1998, p. 286).

Hypothesis H_{21} is also empirically confirmed, although at a lower significance level (ß = .203, p < .01). Openness is positively related to the assimilation capability. The findings highlight that openness is of considerable relevance for the exploitation capability and is also in line with the expectations derived from the literature and innovation, the knowledge-based view, managerial cognition and coevolution in section 3.3.4. It is necessary for knowledge sharing that knowledge transparence and openness enables members both to orient themselves within a company and to access external fields of knowledge, because this fosters the generation of synergies and cooperation, etc., and the company's internal and external resources are used more efficiently (cf. Probst et al. 2010, p. 65). Furthermore, the results of the regression analysis support, from a coevolutionary perspective, the statement by *Cohen & Levinthal* (1994) that openness is a critical factor in industrial competitiveness. It also enables firms both to exploit external knowledge and to predict more accurately the nature of future technological advances (cf. Cohen & Levinthal 1994, p. 227). Therefore, dominant logic is an emergent property of organizations as complex adaptive systems from a managerial cognition perspective (cf. Bettis & Prahalad 1995, p. 14).

Hypothesis H_{22} is not empirically confirmed because the ß of the autonomy construct is negative and non-significant (ß = -.098, n.s.). Thus, the results reject the expectation, based on the literature and the knowledge-based view and coevolution set forth in section 3.3.4, that autonomy is positively related to the exploitation capability. The reason for the negative relationship may be accounted for by an assumption made by *Nonaka* (1994): "By allowing people to act autonomously, the organization may increase the possibility of introducing unexpected opportunities" (Nonaka 1994, p. 18)[153]. The exploitation of opportunities is very important for the exploitation capability, because it denotes a firm's capacity to create something new, but if firms are confronted with too many opportunities, they might lose focus or work inefficiently.

According to a β of .291 at a significance level of .1% (ß = .291, p < .001), support is found for *hypothesis H_{23}*. The finding that learning receptivity is highly relevant for the exploitation capability supports the suggestion that employees who are receptive to learning can set the stage for overcoming barriers to the exploitation of external knowledge. This finding is also in line with

---

[153] *Nonaka* (1994, p. 18), referencing the work of *Cohen et al.* (1972) titled 'A garbage can model of organizational choice'.

the expectations derived from the literature and organizational learning, dynamic capabilities, the knowledge-based view, managerial cognition and coevolution in section 3.3.4. Once again, it is noted in this analysis that learning receptivity is inevitable for the exploitation capability, because exploitation denotes a firm's capacity to create something new by being learning receptive for harvesting and incorporating transformed knowledge into its own operations, which creates *dynamic capabilities*. Employees must be willing to learn new routines, competencies and technologies: they must be willing to change.

Hypothesis H_{24} is not empirically confirmed as well as *hypothesis* H_{23} because the ß of the care construct is negative and non-significant (ß = -.013, n.s.). The findings highlight that care is negatively related to the exploitation capability and is not of considerable relevance for the exploitation capability, which is opposed to the hypothesis generated from the literature and the knowledge-based view and managerial cognition in section 3.3.4. The reason for this negative relationship might be the enhanced courage that employees exhibit towards each other (cf. von Krogh 1998, p. 138). On one side, this courage is very important for employees to give opinions, to propose ideas or to give feedback (cf. Sollberger 2006, p. 127), but on the other hand, it can lead to wantonness, which can have a negative effect on the exploitation capability because the employees may have unrealistic aims, overestimate themselves or fail to take their tasks seriously.

Table 45 gives an overview of the results of the examination of hypothesis H_{19} through H_{24}. The empirical results confirm the expected positive relationship between three of the organizational dimensions, namely, collaboration and the exploitation capability (*hypothesis* H_{20}), openness and the exploitation capability (*hypothesis* H_{21}) and learning receptivity and the exploitation capability (*hypothesis* H_{23}). The hypotheses about the positive relationship between openness and the exploitation capability (*hypothesis* H_{19}), autonomy and exploitation (*hypothesis* H_{22}) and care and the exploitation capability (*hypothesis* H_{24}) are rejected.

Table 45 Overview of the Hypotheses H_{19} to H_{24}[154]

	No.	Hypothesis	Decision
	H_{19}	Trust is *positively* related to the exploitation capability.	Rejected
	H_{20}	Collaboration is *positively* related to the exploitation capability.	Accepted
Exploitation Capability	H_{21}	Openness is *positively* related to the exploitation capability.	Accepted
	H_{22}	Autonomy is *positively* related to the exploitation capability.	Rejected
	H_{23}	Learning receptivity is *positively* related to the exploitation capability.	Accepted
	H_{24}	Care is *positively* related to the exploitation capability.	Rejected

Section 4.9 gives a summary of the regression analysis.

4.9 Summary of the Results

From a business perspective, it is very interesting that it is possible to more accurately differentiate the assumption that it is basically an advantage if employees are positively oriented towards the absorption of external knowledge, because the four regression analyses in the previous sections show that nine of the 24 hypotheses are accepted and 15 are rejected (cf. Table 46). The results show that organizational culture is not a blanket criterion for the success of knowledge absorption (as was announced at the beginning regarding the open and positive organizational culture), but that the individual dimensions of organizational culture are critical.

In detail, the results of the four regression analyses show that several dimensions of organizational culture are positively related to three of the four capabilities of absorptive capacity (cf. Table 46):

---

[154] Author's own table.

4 An Empirical Analysis of the Research Models

- Trust (*hypothesis* H_7), collaboration (*hypothesis* H_8) and openness (*hypothesis* H_9) are positively related to the **assimilation capability**;
- Trust (*hypothesis* H_{13}), collaboration (*hypothesis* H_{14}) and learning receptivity (*hypothesis* H_{17}) are positively related to the **transformation capability**; and
- Collaboration (*hypothesis* H_{20}), openness (*hypothesis* H_{21}) and learning receptivity (*hypothesis* H_{23}) are positively related to the **exploitation capability**.

The results show that beside acquisition, all of the capabilities are positively influenced by the dimensions of organizational culture. The assimilation, transformation and exploitation capabilities are influenced by different dimensions of organizational culture, but the same number of dimensions of organizational culture, namely three (cf. Table 47). Therefore, organizational culture is equally important for the three phases.

Table 46 Summary of the Examination of Hypotheses H_1 through Hypotheses H_{24}[155]

		No.	Hypothesis	Decision
Acquisition Capability		H_1	Trust is *positively* related to the acquisition capability.	Rejected
		H_2	Collaboration is *positively* related to the acquisition capability.	Rejected
		H_3	Openness is *positively* related to the acquisition capability.	Rejected
		H_4	Autonomy is *positively* related to the acquisition capability.	Rejected
		H_5	Learning receptivity is *positively* related to the acquisition capability.	Rejected
		H_6	Care is *positively* related to the acquisition capability.	Rejected
Assimilation Capability		H_7	Trust is *positively* related to the assimilation capability.	Accepted
		H_8	Collaboration is *positively* related to the assimilation capability.	Accepted
		H_9	Openness is *positively* related to the assimilation capability.	Accepted
		H_{10}	Autonomy is *positively* related to the assimilation capability.	Rejected
		H_{11}	Learning receptivity is *positively* related to the assimilation capability.	Rejected
		H_{12}	Care is *positively* related to the assimilation capability.	Rejected
Transformation Capability		H_{13}	Trust is *positively* related to the transformation capability.	Accepted
		H_{14}	Collaboration is *positively* related to the transformation capability.	Accepted
		H_{15}	Openness is *positively* related to the transformation capability.	Rejected
		H_{16}	Autonomy is *positively* related to the transformation capability.	Rejected
		H_{17}	Learning receptivity is *positively* related to the transformation capability.	Accepted
		H_{18}	Care is *positively* related to the transformation capability.	Rejected
Exploitation Capability		H_{19}	Trust is *positively* related to the exploitation capability.	Rejected
		H_{20}	Collaboration is *positively* related to the exploitation capability.	Accepted
		H_{21}	Openness is *positively* related to the exploitation capability.	Accepted
		H_{22}	Autonomy is *positively* related to the exploitation capability.	Rejected
		H_{23}	Learning receptivity is *positively* related to the exploitation capability.	Accepted
		H_{24}	Care is *positively* related to the exploitation capability.	Rejected

With regard to the single dimensions of organizational culture influencing the capabilities of absorptive capacity, the results of the empirical analysis of this

---

[155] Author's own table.

4 An Empirical Analysis of the Research Models

thesis confirm the expectation that dimensions of organizational culture — namely, trust, collaboration, openness and learning receptivity — are positively related to capabilities of the absorptive capacity. Table 47 highlights the following positive relationships:

- **Trust** is positively related to the assimilation capability (*hypotheses* H_7) and the transformation capability (*hypotheses* H_{13}), and therefore to two capabilities of absorptive capacity.
- **Collaboration** is positively related to the assimilation capability (*hypotheses* H_8), the transformation capability (*hypotheses* H_{14}) and the exploitation capability (*hypotheses* H_{20}), and therefore to three capabilities of absorptive capacity.
- **Openness** is positively related to the assimilation capability (*hypotheses* H_9) and the exploitation capability (*hypotheses* H_{21}), and therefore to two capabilities of absorptive capacity.
- **Learning receptivity** is positively related to the transformation capability (*hypotheses* H_{17}) and the exploitation capability (*hypotheses* H_{23}), and therefore to two capabilities of absorptive capacity.

Collaboration has the most positive effect on absorptive capacity because it is positively related to three capabilities of absorptive capacity, followed by trust, openness and learning receptivity, which each have positive relationships to two capabilities of absorptive capacity. Furthermore, the results of the four regression analyses show that one dimensions of organizational culture is negatively related to one of the four capabilities of absorptive capacity: **autonomy** is negatively related to the assimilation capability (*hypotheses* H_{10}; cf. Table 47).

Table 47 Positive Relationships for Each Capability of Absorptive Capacity[156]

Organizational Culture \ ACAP	Trust	Collaboration	Openness	Autonomy	Learning Receptivity	Care	Sum (Capabilities)
Acquisition Capability							0
Assimilation Capability	++	+++	++	-			3+1
Transformation Capability	+++	+++			+		3
Exploitation Capability		+++	++		+++		3
Sum (Dimensions)	2	3	2	(1)	2	0	

With regard to the rejected hypotheses, it is necessary to differentiate between the rejected hypotheses on the one side that only concern the six relationships between the dimensions of organizational culture and the acquisition capability (*hypothesis H_1 - hypothesis H_6*) and the rejected hypotheses on the other hand that concern the relationship between the different dimensions of organizational culture and the assimilation capability (*hypothesis H_{10} - hypothesis H_{12}*), the transformation capability (*hypothesis H_{15}, hypothesis H_{16}, hypothesis H_{18}*) and the exploitation capability (*hypothesis H_{19}, hypothesis H_{22}, hypothesis H_{24}*). The findings related to the relationship between the dimensions of organizational culture and the acquisition capability (*hypothesis H_1 - hypothesis H_6*) are very

---

[156] Author's own table; the * corresponds to the significance level (* $p < .05$, ** $p < .01$, *** $p < .001$) and the algebraic sign corresponds to the positive (+) or negative (-) value of the β: * = +/-; ** = ++/--; *** = +++/---.

4 An Empirical Analysis of the Research Models 215

interesting because they show that organizational culture plays no role in contact with the external environment[157].

The rejected hypotheses related to the relationship between the different dimensions of corporate culture and the capabilities of absorptive capacity cannot be generalized and must be investigated in more detail: six of the remaining 9 rejected hypotheses were rejected because they did not score significantly, and only 3 of the rejected hypotheses were rejected because the direction (positive vs. negative) was incorrectly deduced. However, only the results for *hypothesis H_{10}*, which erroneously assumed a positive relationship, were significant. Therefore, in summary, *hypothesis H_{11}*, *hypothesis H_{12}*, *hypothesis H_{15}*, *hypothesis H_{16}*, *hypothesis H_{18}*, *hypothesis H_{19}*, *hypothesis H_{22}* and *hypothesis H_{24}*, are rejected because they are non-significant[158].

To summarize, the results show that the absorptive capacity of SMEs is influenced by organizational culture. This finding is in accordance with the insights of the literature. Furthermore, it is in accordance with the insights of the literature that the absorptive capacity of SMEs is positively influenced by organizational culture. As the literature identified the generally positive influence of organizational culture on absorptive capacity, with due regard to the theoretical concepts at the beginning of this thesis, a positive relationship between the dimensions of organizational culture and the capabilities of absorptive capacity was assumed and confirmed.

The results of this thesis show that the acquisition capability is the only capability of absorptive capacity that is not influenced by organizational culture. Only during the acquisition of knowledge do firms have contact with their external environments. One reason that the organizational culture has no influence

---

[157] The reasons for the rejection for the hypotheses with regard to the acquisition of external knowledge are discussed in section 4.8.2.1.
[158] The reasons for the rejection for the hypotheses with regard to the assimilation capability are discussed in section 4.8.2.2, with regard to the transformation capability in section 4.8.2.3 and with regard to the exploitation capability in section 4.8.2.4.

on the acquisition capability might be that firms are so harshly confronted with their need for resources to acquire knowledge and their need to take less care of their artifacts, values, and basic underlying assumptions. Another reason might be that there are such strict rules for the acquisition of external knowledge in the scope of interactions with the knowledge sender that it would be impossible for organizational culture to play a role.

The other capabilities — the assimilation capability, the transformation capability and the exploitation capability — are positively influenced by different dimensions of organizational culture. This finding is also in accordance with the insights of the literature, and creates awareness that organizational culture is not, in general, positively related to the capabilities of absorptive capacity, but rather, a detailed differentiation of the several dimensions of organizational culture is necessary to purposefully support external knowledge absorption in a particular organizational culture.

The assimilation capability, the transformation capability and the exploitation capability are all influenced by the same number of dimensions of organizational culture — three. Therefore, organizational culture is equally important for the three capabilities. With regard to the single dimensions of organizational culture, collaboration has the greatest influence on the three capabilities of absorptive capacity and care has no influence on absorptive capacity. Once again, it is important to be aware that organizational culture is not, in general, positively related to the capabilities of absorptive capacity, but rather a detailed differentiation of the several dimensions of organizational culture is necessary to purposefully support the external knowledge absorption in a particular organizational culture.

Following a clear procedure for the development of the model and for the empirical analysis, the realization of a methodologically valid analysis to maximize the consistency and conclusiveness of the results of this thesis were attempted. Limitations concerning the possibility of generalizing the findings, along with starting points for future research, primarily result from the concep-

4 An Empirical Analysis of the Research Models

tualization of the model and the database of this thesis. They are discussed together with further research needs in section 5.

5 Summary, Conclusion and Outlook

This section concludes this thesis with an overview of its findings and contributions to management and research along with its implications for management and research. The first section 5.1 begins with an overall summary of this thesis and its contribution to management and research. The second section 5.2 addresses the implications for management with respect to trust, collaboration, openness and learning receptivity, which are the dimensions of organizational culture that are relevant to the support of absorptive capacity. The third section 5.3 discusses the limitations of the investigation along with further research needs.

5.1 Summary and Contribution to Management and Research

The objective of this thesis is to develop and validate a model that allows an analysis of the relationship between organizational culture and the capabilities of absorptive capacity at the organizational level of SMEs and shows how a knowledge-oriented organizational culture should be designed to support the absorption of external knowledge. The starting point of the analysis is the finding that although the construct of absorptive capacity has received considerable academic attention in recent years, there is still an underdeveloped research area, namely, absorptive capacity in the SME context. Moreover, there has been no investigation of how the several dimensions of organizational culture are related to acquisition capability, assimilation capability, transformation capability and exploitation capability or what courses of action related to organizational culture can be implemented by SMEs to promote the absorption of external knowledge. There is a huge research gap related to the relationship between organizational culture and absorptive capacity and a model that takes into account the several dimensions of organizational culture and the several capabilities of absorptive capacity. Furthermore, there has been no empirical analysis of the several relationships among the dimensions of organizational culture and the capabilities of absorptive capacity.

The results of this thesis include confirmation of its initial hypothesis: Organizational culture is positively related to a company's absorptive capacity. How successfully external knowledge is absorbed depends primarily on the several dimensions of a company's organizational culture.

This investigation of organizational culture as an influencing factor of the SMEs' capacity to absorb external knowledge was divided into four procedural steps, which successively pursued the objective of this thesis:

In the first step, conceptual principles were discussed to develop a model of external knowledge absorption. These principles included the conceptual principles of absorptive capacity and organizational culture. Clarification of the following questions was central to the analysis of the conceptual principles of absorptive capacity and organizational culture: Which conceptual principles describe the research fields related to absorptive capacity and organizational culture, and which working definitions make sense?

For the purposes of discussing the conceptual principles of absorptive capacity, the model of absorptive capacity is explained after a description of the relationship between absorptive capacity and firm performance. The reason for explaining absorptive capacity is that absorptive capacity is an important performance-enhancing lever for SMEs, which typically lack internal resources and capabilities and therefore tend to depend heavily on absorptive capacity because external knowledge absorption is a critical factor for competitiveness. The model of absorptive capacity is characterized by key antecedents, moderators and outcomes of the construct, components and capabilities (acquisition, assimilation, transformation and exploitation) and can be separated into two components, namely, potential absorptive capacity and realized absorptive capacity.

5 Summary, Conclusion and Outlook

For the purpose of discussing the conceptual principles of organizational culture, a definition of organizational culture with its elements and levels is provided, and several approaches of organizational culture are explained. This definition and explanation permits a detailed consideration of organizational culture and those dimensions of organizational culture that play a central role in the absorption of external knowledge. Several approaches of organizational culture are compared with respect to the concept of organizational culture, design of the measurement instrument, dimensions and indicators. Because the literature has identified a generally positive influence of organizational culture on absorptive capacity, with due regard to the theoretical concepts set forth at the beginning of this thesis, a positive relationship between the dimensions of organizational culture and the capabilities of absorptive capacity is assumed. With regard to the context of this thesis, *Sollberger* (2006) was the only one of the listed authors who worked out those dimensions of organizational culture that play a central role in the absorption of external knowledge: trust, collaboration, openness, autonomy, learning receptivity and care. Based on these dimensions, organizational culture has been understood as an approach that defines appropriate observable artifacts, values and basic underlying assumptions and determines what should be of particular relevance to employees. This understanding allows one to describe companies by using the manifestation of the several dimensions of organizational culture. Relying on *Sollberger*'s (2006) manifestation of seven dimensions, various types of organizational cultures at several companies can be formulated. This allows a discussion of the companies' different organizational cultures and their relationship to absorptive capacity.

The second step involves the development of theoretical relationships within the model of external knowledge absorption in the context of learning, innovation, dynamic capabilities, the knowledge-based view, managerial cognition and coevolution. This step addresses the question of how to design a model that shows the relationship between the dimensions of organizational culture and the capabilities of absorptive capacity. The objective of this modeling is to reproduce firms' external knowledge absorption based on their organizational cultures to derive a design concept for an external-knowledge-supporting organizational culture. The modeling was carried out in three substeps:

In the first substep, theories that inform absorptive capacity were introduced to provide a deeper understanding of the concept of absorptive capacity. In their reconceptualization, *Zahra & George* (2002) showed that organizational learning, innovation, dynamic capabilities, the knowledge-based view, managerial cognition and coevolution are the theories that are able to explain the concept of absorptive capacity in its entirety.

In a second step, the parameters of the model that are relevant to identifying the starting points of a cultural design concept for external knowledge absorption, in particular organizational culture contexts, were worked out. Those starting points were absorptive capacity and organizational culture. With regard to absorptive capacity, the four capabilities of absorptive capacity, acquisition, assimilation, transformation and exploitation were explained in detail. With regard to organizational culture, the six dimensions of organizational culture — trust, collaboration, openness, autonomy, learning receptivity and care — were similarly explained. After explaining the theories that inform absorptive capacity and the parameters of the model, the relationships among the dimensions of organizational culture and the capabilities of absorptive capacity were worked out.

An empirical analysis of the research model was part of the third step. To determine which of the relationships among the dimensions of organizational culture and the capabilities of absorptive capacity can be empirically confirmed, an empirical analysis of the research model was conducted in several substeps. First, to carry out the empirical analysis of the research model, SMEs were defined as the object of study following the definition of the *Institute for SME Research Bonn* (2012). Second, the process of data collection was explained. The hypotheses regarding the relationships among the dimensions of organization culture and the capabilities of absorptive capacity that were represented in the model were tested via the quantitative research method of a survey, which

5 Summary, Conclusion and Outlook

allowed for the SMEs' organizational cultures to be systematically captured. Because organizational culture and absorptive capacity were considered as organizational constructs characterized by collectivism, and individual psychological aspects were excluded from the empirical investigation, the questioning of key informants made sense as a data-collection method. To collect the data, all of the variables were operationalized and implemented using a questionnaire. Based on *Zahra & George*'s (2002) working definition of absorptive capacity as a set of routines and processes that enable firms to acquire, assimilate, transform and exploit knowledge (cf. Zahra & George 2002, p. 186), the construct of absorptive capacity was operationalized by its four capabilities — acquisition, assimilation, transformation and exploitation — corresponding to the operationalization of *Zahra & George*'s (2002) four capabilities. Based on the statement by *Sollberger* (2006) that trust, collaboration, openness, autonomy, learning receptivity and care are the values of a knowledge-friendly organizational culture (cf. Sollberger 2006, p. 119), the construct of organizational culture was operationalized using its six dimensions — trust, collaboration, openness, autonomy, learning receptivity and care — corresponding to the *Sollberger*'s (2006) operationalization of those dimensions.

After the data were collected via the questionnaire, which was distributed, for example, through channels such as institutional websites and newsletters, databases, publishing in XING groups etc., they were described. Altogether, 472 questionnaires could be analyzed. The distribution of industry affiliation and firm size is heterogeneous. The firms' sales are less heterogenic because three categories dominate: 23.9% have less than €500,000 sales, 23.3% have between €500,000 and €1 million sales and 36.4% have between €2 million and €9 million sales. Most of the respondents worked in management or marketing and sales and had managerial functions. Following the description of the sample, the data were analyzed via a descriptive analysis, a reliability analysis, a factor analysis and a regression analysis. Finally, the results of the empirical analysis were summarized.

The results of the empirical analysis of the model of external knowledge absorption lead to very important contributions to management and research, which are presented in the following sections.

5.1.1 Contributions to Management

The results show that an SME's absorptive capacity is influenced by organizational culture. For management, this requires awareness that organizational culture is an influencing factor of the capacity of SMEs to absorb external knowledge and develop strategies for how to support their capacity to absorb external knowledge through a suitable organizational culture. Implications for management are given in section 5.2.

In addition to the influence of organizational culture on SMEs' absorptive capacity, the results of this thesis show that acquisition capability is the only capability of absorptive capacity that is not influenced by organizational culture. For management, this means that their own organizational culture plays no role in acquisition capability and contact with the external environment. One reason that organizational culture has no influence on acquisition capability might be that firms are so heavily confronted with the facts both that they need resources to acquire knowledge and that they should spend less time on their artifacts, values, and basic underlying assumptions. Another reason for the lack of organizational culture's influence on acquisition capacity might be that there are such strict rules for the acquisition of external knowledge in the scope of interactions with a knowledge sender that it is not possible for organizational culture to play a role. Management should probe whether this is really the reason for that organizational culture does not play enough of a role in acquisition capability to be able to control an acquisition with these rules.

In addition to acquisition capability, a firm's other capabilities — assimilation, transformation and exploitation — are positively influenced by different dimensions of organizational culture. This finding is very important to

5 Summary, Conclusion and Outlook 225

management because it should be aware that in general organizational culture is not positively related to the capabilities of absorptive capacity; instead, a detailed differentiation of the several dimensions of organizational culture is necessary to purposefully support external knowledge absorption in a particular organizational culture.

The results show that in addition to acquisition capability, all of a firm's capabilities are positively influenced by dimensions of organizational culture: Trust, collaboration and openness are positively related to assimilation capability; trust, collaboration and learning receptivity are positively related to transformation capability; and collaboration, openness and learning receptivity are positively related to exploitation capability. Therefore, assimilation capability, transformation capability and exploitation capability are influenced by different dimensions of organizational culture, but the same number of dimensions — three — of organizational culture. This thesis's contribution, for management, is the finding that organizational culture is equally important to the three capabilities of absorptive capacity. Management can collectively support the different capabilities of absorptive capacity if it considers the positive relationship among the dimensions of organizational culture and the capabilities of absorptive capacity.

The single dimensions of organizational culture that influence the capabilities of absorptive capacity — trust, collaboration, openness and learning receptivity — are positively related to capabilities of the absorptive capacity. Trust is positively related to assimilation and transformation capabilities; collaboration is positively related to assimilation, transformation and exploitation capabilities; openness is positively related to assimilation and exploitation capabilities; and learning receptivity is positively related to transformation and exploitation capabilities. Collaboration has the most positive effect on absorptive capacity because it is positively related to three capabilities of absorptive capacity. Trust, openness and learning receptivity have the second-most positive effects on absorptive capacity because each of these dimensions has positive relationships with two capabilities of absorptive capacity[159]. Once again, it is important for management to be aware that organizational culture is generally not positively related to the capabilities of absorptive capacity; instead, a detailed differentiation of the several dimensions of organizational culture is necessary to purposefully support external knowledge absorption in a particular organizational culture.

In summary, the results show that organizational culture helps to solve the problem of SMEs' external knowledge absorption. In their efforts to maintain important positions in the marketplace and to remain globally competitive in times of rough market conditions, SMEs are challenged by their lack of knowledge resources. This challenge is especially important because knowledge is a prerequisite for innovation, which enables firms to generate long-term competitive advantages. Additional challenges that confirm the vulnerability of SMEs include both the major liability of deficiencies in internal financial resources and a lack of both human resources and specialized knowledge. In summary, SMEs have a lack of the resources that are necessary to generate long-term competitive advantages. Therefore, resources are SMEs' primary challenge. To address this resource challenge, firms complement their knowledge base by acquiring external knowledge that they normally cannot develop independently. To explore the external knowledge, SMEs need absorptive capacity.

This thesis confirms that organizational culture is an influencing factor of absorptive capacity, and therefore, it is very important for competitiveness — especially for SMEs — because the establishment of a knowledge-friendly cul-

---

[159] Furthermore the results of the four regression analyses show that one dimension of organizational culture is negatively related to one of the four capabilities of absorptive capacity: Autonomy is negatively related to the assimilation capability, but autonomy concerns only one capability of absorptive capacity (the assimilation capability) and the ß is so small (ß = -.122, p < .05) that autonomy was not considered in the design concept as an influencing factor.

ture is a possible method by which SMEs can distinguish themselves in the market and take advantage of their specific competitive advantages. In addition to providing this advantage of SME-specific competitiveness, a strongly knowledge-friendly organizational culture is highly important to an SME's long-term survival because processes in dynamic, uncertain environments can be better coordinated through implicit values and norms better than through structural coordination instruments. The reason why these processes can be coordinated through implicit values and norms is that organizational culture influences people's knowledge-related behaviors: it strongly influences the determination of which knowledge is appropriate to share, with whom and when. The results of this thesis show which implicit values and norms play an important role for absorptive capacity and for the single capabilities. Trust, collaboration, openness and learning receptivity are key to influencing people's knowledge-related behaviors while they absorb external knowledge. It is particularly important that SMEs exploit this potential of organizational culture because due to their resource poverty, small- and medium-size companies usually do not have the resources to comprehensively establish explicit structural coordination instruments or motivation rules. In addition, SMEs should not give up their advantage of strategic flexibility, which can be seen in their less-bureaucratic structures.

5.1.2 Contributions to Research

Overall, this dissertation successfully addresses the two previously mentioned problem areas in the research:

The first problem area includes unresolved questions with respect to the parameters of a model of SMEs' external knowledge absorption. Previously, it was unclear which dimensions of the prevailing organizational culture influence the several capabilities of SMEs' absorptive capacity; moreover, it was also unclear which indicators could be used to measure these dimensions of a knowledge-friendly organizational culture. It can be determined that trust, collaboration, openness, autonomy, learning receptivity and care, as dimensions of a knowledge-friendly organizational culture, influence the several capabilities of SMEs' absorptive capacity.

The second problem area includes unresolved questions with respect to the relationship among the dimensions of organizational culture and the capabilities of absorptive capacity. This understanding was important for designing a model that permits an assessment of organizational culture's role in ensuring the absorption of external knowledge.

This thesis presents a design of a model that permits an assessment of the role of organizational culture in ensuring the absorption of external knowledge. It is possible both to confirm positive relationships among the dimensions of organizational culture and the capabilities of absorptive capacity using this novel model of external knowledge absorption and to identify principles for a design concept for an external-absorption-supporting organizational culture and therefore, to close the research gap to some extent[160].

With regard to theoretical relevance, the following points can be made:

For the theoretical foundation of this thesis, theories informing absorptive capacity — namely, organizational learning, innovation, dynamic capabilities, the knowledge-based view, managerial cognition and coevolution — proved to be of particular relevance. The knowledge-based view provides the most comprehensive theoretical embedding of the research model set out in this thesis, which considers knowledge as a firm's most important resource because it is the main determinant of competitive advantage. This view influences the

---

[160] The positive relationships among the dimensions of organizational culture and capabilities of absorptive capacity that cannot be confirmed lead to a new research gap with regard to the theories informing absorptive capacity, because those theories are insufficient to explain and predict how the dimensions of organizational culture influence the capabilities of absorptive capacity. Section 5.3 explains in detail that further research is necessary.

5 Summary, Conclusion and Outlook

relevance of absorptive capacity because absorptive capacity is the key to developing and increasing a firm's knowledge base. Furthermore, the knowledge-based view demonstrates the importance of the resource of knowledge for each of the four capabilities of absorptive capacity. It further provides the basis for identifying the relationships among the dimensions of organizational culture and the capabilities of absorptive capacity to find out how a how a knowledge-oriented organizational culture should be designed to support the absorption of external knowledge.

Most of the previous studies in the field of knowledge transfer address absorptive capacity in only a rudimentary way. However, as the results of this thesis illustrate, absorptive capacity is informed by organizational learning, innovation, dynamic capabilities, the knowledge-based view, managerial cognition and coevolution, which together represent a fundamental theoretical foundation for this research area. Therefore, this study offers a comprehensive insight into absorptive capacity.

In summary, this dissertation makes a considerable contribution to the research area of organizational culture and absorptive capacity, from a theoretical standpoint. The attempt to substantially specify the phenomenon of organizational culture to enable both an empirical investigation and a practical application of organizational culture as an absorptive-capacity-influencing factor was successful. For the analysis of a company's organizational culture, it is therefore essential to substantively specify the organizational culture within a company to operationalize it. Organizational culture is not, as assumed at the beginning of this thesis, generally positively related to the capabilities of absorptive capacity, but rather is a detailed differentiation of the several dimensions of organizational culture necessary to reasonably use principles of the design concept to support external knowledge absorption in a particular organizational culture. This dissertation's major contribution to theory is that the blanket image of organizational culture is differentiated and that the consequences for how a knowledge-oriented organizational culture should be designed to support the absorption of external knowledge are drawn. Furthermore, the theoretical foundation of absorptive capacity is advanced because how the several theories informing absorptive capacity — organizational learning, innovation, dynamic capabilities, knowledge-based view, managerial cognition and coevolution — affect each of the four capabilities of absorptive capacity is explicitly worked out.

5.2 Implications for Management

The findings of this study offer new and valuable insights on organizational culture as an influencing factor of SMEs' capacity to absorb external knowledge. Based on an extensive literature review and the theories informing absorptive capacity, a comprehensive model of external knowledge absorption is presented. This model can help managers to become aware of dimensions of organizational culture that influence the capability to assimilate, transform and exploit knowledge: trust, collaboration, openness and learning receptivity.

Because the model is used to both describe the absorption of external knowledge and explain the influence of organizational culture so as to derive starting points for organizational-culture-driven control of external knowledge absorption, managers can learn methods of addressing the challenges of external knowledge absorption using a knowledge-friendly organizational culture. The results of this thesis provide important managerial implications concerning the successful absorption of external knowledge.

To develop and design an organizational culture within the context of the topic of this thesis, management must know the relationship among the dimensions of organizational culture and the capabilities of absorptive capacity: trust, collaboration, openness and learning receptivity are part of a knowledge-friendly, absorptive-capacity-supporting organizational culture (cf. Figure 25). To add detail, trust is positively related to assimilation and transformation capabilities, collaboration is positively related to assimilation, transformation and fexploitation capabilities, openness is positively related to assimilation and

5 Summary, Conclusion and Outlook

exploitation capabilities, and learning receptivity is positively related to transformation and exploitation capabilities.

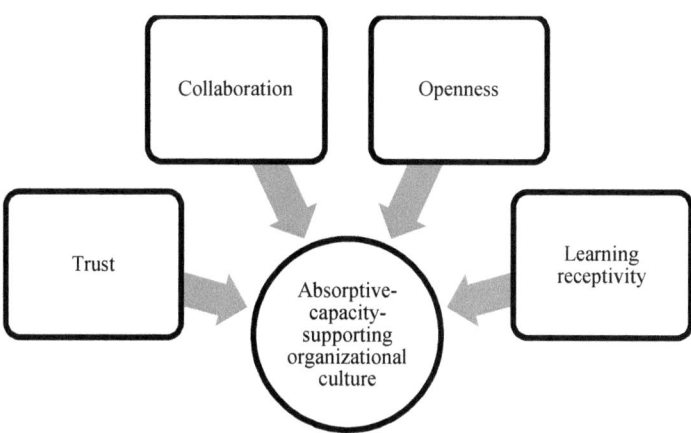

Figure 25 An Absorptive-Capacity-Supporting Organizational Culture[161]

The following implications include recommendations for business practices that refer to a knowledge-oriented organizational culture that influences the capabilities of absorptive capacity. The implications of designing a knowledge-oriented organizational culture to support the absorption of external knowledge, described subsequently, are structured along the several important dimensions of organizational culture: trust, collaboration, openness and learning receptivity.

The management implications of creating a trustful, collaborative, open and learning-receptive culture are also provided. These implications begin with implications for a trustful organizational culture in section 5.2.1.

---

[161] Author's own figure,

5.2.1 Implications for a Trustful Organizational Culture

As shown by the literature review and the empirical analysis, trust is considered to be one of the most important enablers of the assimilation and transformation of external knowledge. With regard to a trustful organizational culture, *Davenport & Prusak* (1998) emphasize the importance of the existence of trust: "Without trust, knowledge initiatives will fail, regardless of how thoroughly they are supported by technology and rhetoric and even if the survival of the organization depends on effective knowledge transfer" (Davenport & Prusak 1998, p. 34). *Pemberton & Stonehouse* (2002) support the importance of trust: "A degree of trust must be fostered throughout the organization, encouraging personal responsibility for knowledge dissemination" (Pemberton & Stonehouse 2002, p. 85).

Degree of trust depends on several conditions. A firm in which trust is high can be described as follows: "Organizational participants share certain ends or values; bear towards each other a diffuse sense of long-term obligations; offer each other spontaneous support without narrowly calculating the cost or anticipating any short-term reciprocation; communicate honestly and freely; are ready to repose their fortunes in each other's hands; and give each other the benefit of any doubt that may arise with respect to goodwill or motivation" (Armstrong 2006, p. 221)[162]. This description presents a picture of a firm with an organizational culture that facilitates trust. To create organizational culture that facilitates trust, *Davenport & Prusak* (1998) have worked out three ways in which trust must be established for the knowledge market to operate in an organization: Trust must be visible and ubiquitous, and trustworthiness must start at the top (cf. Davenport & Prusak 1998, p. 34).

Trust must be visible. An organizational culture in which trust plays an important role is characterized by the feeling of the employees that they realize profit from absorbing external knowledge. The employees trust that the compa-

---

[162] *Armstrong* (2006, p. 221), referencing, *Fox* (1973).

ny values their investment of resources into absorbing knowledge: If people give their knowledge to the firm and trust the company, they are confirmed in their behavior by the feelings that they are doing the right thing and that offering each other trustful support is conductive to their own and the company's absorptive capacity and competitive advantage. Incentives can award people for their behavior and motivate them to live corresponding to a trustful philosophy. Furthermore, incentives can make the success of the absorption of external knowledge visible, if management communicates that success and that success was due to trust. This communication can encourage the discussion of knowledge which makes trust visible when people talk about it. Furthermore, this communication is very important for the assimilation of knowledge to help with analyzing, processing, interpreting, and understanding acquired knowledge, and for exploiting that knowledge to refine, extend, and leverage existing routines and processes or to create new ones by incorporating knowledge into a firm's own routines and processes. To make trust more visible, it is important to include trust in a firm's corporate principles so that everybody knows that trust is requested. In addition to the visibility of trust, the visibility of security is required: in other words, when employees share incorrect knowledge or otherwise fail, they must be secure in the fact that they will not be shamed. This visibility of security through trust encourages a climate that is conducive to the assimilation and exploitation of knowledge: it alleviates the fear of risk and uncertainty that play a big role in both capabilities when employees' processes and routines are challenged. It must always be conveyed to employees that it is important to build trust for the absorption of external knowledge. With respect to this dimension of organizational culture — especially the assimilation and transformation capabilities — it must always be clear that building absorptive capacity is for the benefit of the firm.

Trust must be ubiquitous. Untrustworthiness can influence the symmetry and efficiency of the internal knowledge market: "If part of the internal knowledge market is untrustworthy, the market becomes asymmetric and less efficient" (Davenport & Prusak 1998, p. 34). Therefore, trust must be present throughout an entire company to enable the construction of a company-holistic understanding that each individual must behave trustfully to ensure the company's appropriate absorptive capacity, which is based on the firm's knowledge. To avoid isolated applications, all employees must receive basic training so that they share certain goals or values. Employees should be taught that it is important to have the appropriate capabilities of absorptive capacity for themselves and thus, comprehensively, for the entire company. Absorptive capacity is not only important in departments such as marketing, which are very close to the market and interface with it, or in departments where the danger of information asymmetries is larger. Rather, all departments must establish trust to ensure their firm's absorptive capacity. To foster trustworthiness, a policy of transparency must be implemented, intentions and reasons for decisions must be communicated and mutual expectations must be agreed, particularly during the assimilation and transformation of knowledge because trust is positively related to assimilation and transformation capabilities.

Trustworthiness must start at the top. Trust tends to flow downward through organizations because managers can often define a firm's observable artifacts, values and basic underlying assumptions: "If top managers are trustworthy, trust will seep through and come to characterize the whole firm. If they cynically exploit others' knowledge for personal gain, distrust will propagate throughout the company. Their values become known to the firm through signals, signs, and symbols" (Davenport & Prusak 1998, p. 34 f.). This passage stresses that it is not possible to manage trust but that trust is an outcome of preconditions created by the management because trust is created and maintained through management's behavior. In other words, management must behave as role models and set an example of trustworthiness. Management behavior is most likely to engender trust when employees believe that management means what it says, is honest with employees, keeps its word and practices what it preaches (cf. Armstrong 2006, p. 221). "More specifically, trust will be developed if management acts fairly, equitably and consistently, if a policy of transparency is implemented, if intentions and the reasons for proposals or decisions are communicated both to employees generally and to individuals, if there is full involvement in developing HR processes, and if mutual expectations are agreed through performance management" (Armstrong 2006, p. 222)[163]. With regard to top management, it is furthermore important to absorb strategically important knowledge. It is very important that top management, which is responsible for a firm's strategy, absorbs knowledge that other employees might not be able to absorb because they do not have the strategic knowledge base that is necessary for the absorption of strategic knowledge. Therefore, the transformation of external knowledge should happen top-down because it is necessary to have corresponding strategic knowledge to develop and refine those routines that facilitate combining existing knowledge with acquired and assimilated knowledge for future strategic use. Nevertheless, management should never forget to take feedback from the bottom up[164].

5.2.2 Implications for a Collaborative Organizational Culture

Collaboration is very important for nearly all absorption of external knowledge. The literature review and the empirical analysis show that when collaboration is anchored in the organizational culture, assimilation, transformation and exploitation capabilities are positively influenced.

Anchoring collaboration in the organizational culture can be accomplished by taking care of employees' organizational socialization. Organizational socialization has been defined as "the process through which individually acquired the knowledge, skills, attitudes, and behaviors required to a new role" (Wanberg 2012, p. 17). It is important for everybody in the company who moves into a new role: "Organizational socialization may pertain to a new employee beginning work at an organization (i.e., an organizational newcomer), or to an individual moving into a new role within the organization (i.e., a promo-

---

[163] 'HR' is an abbreviation of 'human resources'.
[164] This corresponds to the idea of the middle-up-down management model by *Nonaka & Takeuchi* (1995), which attempts to combine the top-down and bottom-up approaches to organizational management.

tion, lateral transfer, expatriate assignment, etc.)" (Wanberg 2012, p. 17). With respect to employees generally, organizational socialization plays an important role in establishing a collaborative organizational culture. The organizational socialization process creates the basics for later collaboration during the absorption of external knowledge and if employees are differently socialized — e.g., they are new to the firm or to a particular department — it is very important for future collaboration that they be socialized in the same collaborative way. Management can actively influence the socialization process using several measures: providing mentors, consciously fostering teambuilding and creating an infrastructure that supports collaboration.

With regard to mentoring, it is particularly appropriate to communicate cultural values based on visible and audible behavior patterns such as a company's routines and processes (cf. Spieth 2006, p. 347). These routines and processes can be taught through mentoring. *Armstrong* (2006) gives the following definition of mentoring: "Mentoring is the process of using specially selected and trained individuals to provide guidance, pragmatic advice and continuing support, which will help the person or persons allocated to them to learn and develop" (Armstrong 2006, p. 569). Mentors conduct several tasks while mentoring. For example, mentors provide people with guidance as to how to acquire necessary knowledge and skills, advice on dealing with administrative, technical or people problems and coaching (cf. Armstrong 2006, p. 569). Mentors help people to obtain the competences to absorb knowledge.

Another measure that influences the socialization process is team building. Teams can be regarded as a predestined framework to foster effective collaboration: *Spieth* (2009) highlights that team structures automatically promote communication within a company, as personal conversations represent a fundamental basis within teams for graduating joint tasks (cf. Spieth 2009, p. 348). In-team collaboration can be intensified by team-building activities (e.g., Maddux & Wingfield 2003, Dyer, Dyer & Dyer 2003, Quick 1992, Parker & Kropp

1992). An example of team building activities is the activity 'Really... But I Thought' (cf. Parker & Kropp 1992, p. 25)[165]. With regard to the composition of the team, it is very important that it contains all of the capabilities of absorptive capacity. These capabilities can be distributed among different team members, all of whom should be together.

In addition to team-building, a company should support the possibility of informal collaboration by creating an infrastructure that supports collaboration, which means providing a spatially open and communicative atmosphere (e.g., open-space offices, desks in grouped arrangements, etc.) because such an infrastructure allows employees to conduct face-to-face exchanges while working (cf. Spieth 2006, p. 349). In addition, the establishment of forums, chat rooms, coffee corners, smoking spaces, cafeterias, etc. provides platforms for regular knowledge exchange (cf. Glückstein 2000, p. 325). Knowledge exchange is not exclusively focused on company issues, but the established values of the actors themselves are transferred into business routines and processes (cf. Spieth 2006, p. 349).

5.2.3 Implications for an Open Organizational Culture

As a dimension of organizational culture, openness can find expression through interest in issues that are beyond an employee's immediate task area, the receptiveness to actively take up ideas and suggestions from outside, the willingness to actively obtain suggestions from other stakeholders while providing them with learning opportunities, the constructive engagement with opposing opinions, the willingness to share knowledge and the ability to address stress (cf. Armutat, Krause, Linde, Rump, Striening & Weidmann 2002, p. 139). It encourages new and innovative ideas and risk-taking (cf. Hurley & Hult 1998, p.

---

[165] Altogether, 50 team-building activities can be found in *Parker & Kropp* (1992). For each activity, the purpose, group size, time, physical setting, materials, process and variations are explained.

46 f.). This encouragement sends signals to employees that their creativity and engagement are valued, which in turn encourages them to care about innovation for their firm's competitive advantage. Furthermore, openness increases the possibility that greater attention to external knowledge will no longer be confronted with resistance from employees. Interactions between employees, each of whom possess different knowledge structures, will augment their capacity for making novel linkages and associations beyond what any one individual can achieve — assuming a sufficient level of knowledge overlap to ensure effective communication (cf. Cohen & Levinthal 1990, p. 133). Every employee must be willing to actively and constructively address these linkages and associations.

There are three design elements of an open organizational culture: Everything must be generally accessible. Everybody must be able to participate. Everything should have a community-character with few hierarchies. Because openness is positively related to assimilation and exploitation capabilities, these design elements are important for assimilating and exploiting external knowledge.

It is very important that employees feel that everything is generally accessible. For that to be the case, a firm should avoid values in its organizational culture such as power distance, uncertainty avoidance and individualism, which trigger employee behavior that counteracts the sharing and absorption of knowledge. As soon as knowledge is assimilated or finally exploited it should be available to all employees. Measurements for the management of internal knowledge can be so-called knowledge centers, for example, knowledge maps, idea contests, knowledge broker, quality circle, etc. (cf. Glückstein 2000, p. 325)[166]. They help both to spread knowledge in a company and to make that knowledge generally accessible.

It is very important that employees believe that everybody is able to participate. To nurture that belief, it is very important that employees have a

---

[166] *Glückstein* (2000), referencing *ILOI* (1997).

5 Summary, Conclusion and Outlook

positive attitude towards openness. It is important to avoid any belief on the part of employees that they possess a monopoly on knowledge in their area of specialization and that they need not seriously consider the possibility that outsiders might produce important, new knowledge relevant to the company (cf. Katz & Allen 1982, p. 7). This attitude, which is called NIH syndrome, plays an important role with regard to openness, as it refers to a negative attitude towards openness. To overcome NIH syndrome, firms call on specialized, boundary-spanning professionals, who are so-called 'gatekeepers' (cf. Katz & Allen 1982, p. 16, cf. Katz & Tushman 1981, p. 103 ff.). *Katz & Allen* (1982) define gatekeepers "as those key R&D professionals who are both high internal and external communicators and who are also able to effectively transfer external ideas and information into their project groups" (Katz & Allen 1982, p. 16). *Allen & Cohen* (1969) define gatekeepers in more detail as "individuals who occupy key positions in the communication network of the laboratory; that is, those to whom others in the laboratory most frequently turn for technical advice and consultation, and who will show more contact with technical activity outside of the laboratory: a. They themselves will be better acquainted than others in the laboratory with such formal media as the scientific and technological literature. b. They will maintain a greater degree of informal contact with members of the scientific and technological community outside of their own laboratory" (Allen & Cohen 1969, p. 13). Following this definition, gatekeepers are persons who are strongly connected to both internal colleagues and external sources of information (cf. Katz & Trushman 1981, p. 103). The role of a gatekeeper could be assigned either to members of the executive board or a firm's executive employees who, additionally, can develop an appropriate organizational structure, culture and decision-making process for the successful absorption of external knowledge.

In addition to making employees feel that everybody can participate, it is very important that the employees feel that the entire organization has a community-character and that there are few hierarchies. To achieve that feeling, an organization must prevent employees from adopting an 'it's not my job' attitude. The Not-My-Job-ers "express their negativity by refusing to do any task, no matter how simple, if they decide it is not part of their job responsibilities. It is often their way of getting back at their colleagues, their managers, or the organization itself because of their unhappiness with how they are being treated" (Topchick 2001, p. 28). This 'it's not my job' attitude can be avoided by communicating an organization's community-character. If the Not-My-Jobers find training and development opportunities and are involved in a firm's entire working process, they will seek growth and advancement instead of losing their enthusiasm for work and trying to do as little as possible (cf. Topchick 2001, p. 29).

5.2.4 Implications for a Learning-Receptive Organizational Culture

Many companies do not take the opportunity to analyze past mistakes and learn from them: "Successes are proclaimed, but failures are usually buried quickly and unceremoniously without review, with a corresponding loss of potential lessons" (Hatten & Rosenthal 2002). For the long-term success of a company, it is crucial to analyze past mistakes and learn from them, as the literature review and the empirical analysis showed with respect to the transformation and exploitation capabilities. Therefore, management must create an organizational culture of learning receptivity by giving employees tips on how to address both errors and ideas.

According to *Armutat et al.* (2002), learning receptivity is manifested in several ways: "Changes in behavior with regard to the timing; acceptance of external pulses and their influx into the own acting; drawing own conclusions from the results of the actions respectively taking these experiences into account in new situations; requesting feedback on own behavior; open discussion of errors and usage of them as a starting point for improvement" (Armutat et al.

5 Summary, Conclusion and Outlook 241

2002, p. 39 f.; own formatting)[167]. These characteristic manifestations of learning receptivity relate to the absorption of external knowledge as follows:

Changes in behavior related to the timing of the transformation and exploitation of external knowledge play a very important role because employees have learned that it is necessary to develop and refine routines that facilitate combining existing knowledge with both acquired and assimilated knowledge for future use. It is also necessary to refine, extend, and leverage existing routines, competencies and technologies — or to create new ones — by incorporating acquired and transformed knowledge into a firm's operations each time that new knowledge is acquired and assimilated. Learning receptivity is especially important for the transformation and exploitation capabilities because both capabilities build up realized absorptive capacity. Potential and realized absorptive capacity have different but complementary roles in that they coexist at all times and fulfill a necessary but insufficient condition to improve a firm's competitive advantage. Although firms cannot possibly exploit knowledge without first acquiring it, the capability of transforming and exploiting knowledge for profit is essential after successful knowledge acquisition and assimilation because the mere existence of a high-potential absorptive capacity does not necessarily imply enhanced performance. Employees must be learning-receptive and willing to transform and exploit all newly acquired and assimilated knowledge to enhance performance. This transformation and exploitation must happen as much as possible because to stay competitive, SMEs must be able to quickly and flexibly respond to market changes.

In addition to changes in behavior related to timing, the acceptance of external knowledge and the development and refinement of routines and processes that facilitate the combining of existing and external knowledge for future use are essential for the transformation and exploitation of external knowledge. Without the receptivity to learn to address new knowledge, firms

---

[167] Translated by the author.

are not able to generate competitive advantage because they must accept the new knowledge and combine it with existing knowledge. Furthermore, drawing one's own conclusions from the results of knowledge exploitation that takes these experiences into account in new situations is essential for the success of the refinement, extension and leveraging of existing routines, competencies and technologies or for the creation of new ones by incorporating transformed knowledge into a firm's own operations. Requesting feedback on one's own behavior during the refinement, extending and leveraging of existing routines, competencies and technologies helps one to draw his or her own conclusions from the results of the transformation and exploitation of knowledge. Likewise, openly discussing errors helps in locating a starting point for improving the exploitation of external knowledge.

5.3 Implications for Research

The overall objective of this thesis is to develop and validate a model that allows an analysis of the relationship between organizational culture and the capabilities of absorptive capacity at the organizational level of SMEs and shows how a knowledge-oriented organizational culture should be designed to support the absorption of external knowledge. This dissertation targets the workup of the identified problem areas and intends to render manageable the complex system of organizational culture and the complexity of its dimensions and relationships related to the capabilities of absorptive capacity manageable and — with respect to designs of the absorption of external knowledge — available. This thesis contributes to the scientific explanation of the phenomenon of organizational culture, the influence of organizational culture on the absorption of external knowledge and the controllability of organizational culture to support external knowledge absorption. The focus of this thesis is on the theoretical and empirical analysis of this unexplored relationship between the dimensions of an organizational culture and the capabilities of its absorptive capacity.

5 Summary, Conclusion and Outlook

Following a clear procedure for both developing the model and the empirical analysis, this thesis attempts to realize a methodologically valid analysis to maximize the consistency and conclusiveness of its results. However, there are limitations related to the possibility of generalizing the findings, along with starting points for future research, which primarily result from the conceptualization of the model and from the data:

First, the results of this dissertation are derived from analyzing a sample of German SMEs. Future research may investigate whether similar results concerning the relationship between organizational culture and absorptive capacity can be found for a sample of SMEs of other nations. Countries and their organizational culture are often considerably different, so that country-specific adaptations of a design concept for an external-knowledge-absorption-supporting organizational culture will be required. Consequently, it is assumed that the findings of this thesis, with its focus on Germany, cannot be generalized to all other nations, and further research seems advisable. Furthermore, future research may investigate whether similar results concerning the relationship between organizational culture and absorptive capacity can be found for a sample of large firms in Germany or other nations, as large firms and their organizational culture often considerably differ so that specific adaptations of a design concept are required for an external-knowledge-absorption-supporting organizational culture.

Second, future research may extend the theories informing absorptive capacity. Organizational learning, innovation, managerial cognition, the knowledge-based view, dynamic capabilities and coevolution are insufficient to explain and predict how dimensions of organizational culture influence the capabilities of absorptive capacity. If a theory was used to derive a plausible hypothesis, a rejection of the hypothesis is less attributable to its lack of plausibility than it is to the fact that the theory — particularly with regard to the acquisition — is not able to explain the effect. Thus, it is necessary to consider what this means for theory and what further recommendations can be made for developing theory.

Third, this dissertation exclusively examined relationships between organizational culture and absorptive capacity. Future research may investigate whether this dissertation's findings can also be transferred to the desorptive capacity-context. For instance, it may be interesting to investigate whether the ability to externally exploit knowledge is influenced by organizational culture.

Fourth, future research may focus on one industrial sector. Limitations related to the possibility to generalize this dissertation's findings mainly result from its database because the distribution of the industry affiliation is very heterogeneous. A resampling of the empirical study, particularly in the context of an industrial sector-specific analysis, can provide insights as to whether single industrial sectors have sector-specific forms of organizational culture and absorptive capacity.

Fifth, further research on extension and modification of the theoretical foundation of the model of external knowledge absorption in a particular organizational context would be desirable. This dissertation focuses on the model's essential foundation of theories that inform absorptive capacity, namely organizational learning, innovation, dynamic capabilities, the knowledge-based view, managerial cognition and coevolution. In further research, particularly psychological research, theories must be forced. These theories focus more on the behavioral aspect of employees under consideration in an organizational culture and may provide further insight and implications for practice.

Overall, it can be stated that this dissertation has successfully addressed the research question related to the relationship between organizational culture and absorptive capacity. It also demonstrates the relevance of organizational culture to the successful absorption of external knowledge, and the importance of having management implement a design concept for an external-knowledge-absorption-supporting organizational culture. In the beginning, one could make very interesting findings for the defined research gap, although the theories informing absorptive capacity are insufficient to explain and predict several relationships among dimensions of organizational culture and the capabilities of

5 Summary, Conclusion and Outlook

absorptive capacity. Nevertheless, this relationship must be reviewed frequently — from the perspectives of both science and practice — because on the one hand, organizational culture will never be manifested definitively because it is exposed to a constant change, and on the other hand, the requirements for a successful external knowledge absorption will increase significantly as many SMEs are confronted with rough market conditions and must be able to quickly and flexibly respond to market changes to stay competitive.

Literature

Abrams, L. C., Cross, R., Lesser, E. & Levin, D. Z. (2003). Nurturing interpersonal trust in knowledge-sharing networks, *Academy of Management Executive* **17**(4): 64-77.

Ahmed, P. K., Kok, L. K. & Loh, A. Y. E. (2002). *Learning Through Knowledge Management*, Butterworth Heinemann, Oxford.

Ahuja, G., & Katila, R. 2001. Technological acquisitions and the innovation performance of acquiring firms: A longitudinal study, *Strategic Management Journal* **22**: 197-220.

Alavi, M., Kayworth, T. R. & Leidner, D. E. (2006). An empirical examination of the influence of organizational culture on knowledge management practices, *Journal of Management Information Systems* **22**(3): 191-224.

Albers, S. (2010). Configurations of alliance governance systems, *Schmalenbach Business Review* **62**: 204-233.

Allee, V. (1997). The *Knowledge Evolution. Building Organizational Intelligence*, Butterworth Heinemann, Oxford.

Allee, V. (2001). *12 Principles of Knowledge Management*, American Society for Training and Development, Alexandria, Virginia.

Allen, R. F. & Dyer, F. J. (1980). A tool for tapping the organizational unconscious, *Personnel Journal: The magazine of industrial relations and personnel management* **59**(3): 192-198.

Allen, T. J. & Cohen, S. I. (1969). Information flow in research and development laboratories, *Administrative Science Quarterly* **14**(1): 12-19.

Almeida, P. & Kogut, B. (1999). Localization of knowledge and the mobility of engineers in regional networks, *Management Science* **45**(7): 905-917.

Anakwe, U. P., Kessler, E. H. & Christensen, E. W. (1999). Distance learning and cultural diversity: potential users' perspective", *International Journal of Organizational Analysis* **7**(3): 224-243.

Andrews, K. M. & Delahaye, B. L. (2000). Influences on knowledge processes in organizational learning: The psychological filter, *Journal of Management Studies* **37**(6): 797-810.

Appelbaum, S. H. & Gallagher, J. (2000). The competitive advantage of organizational learning, *Journal of Workplace Learning* **12**(2): 40-56.

Appleyard, M. M. (1996). How does knowledge flow? Interfirm patterns in the semiconductor industry, *Strategic Management Journal* **17**(Winter Special Issue): 137-154.

Argote, L., Ingram, P., Levine, J. M. & Moreland, R. L. (2000). Knowledge transfer in organizations: Learning from the experience of others, *Organizational Behavior and Human Decision Processes* **82**(1): 1-8.

Argote, L., McEvily, B. & Reagans, R. (2003a). Introduction to the special issue on managing knowledge in organizations: Creating, retaining, and transferring knowledge, *Management Science* **49**(4): v-viii.

Argote, L., McEvily, B. & Reagans, R. (2003b). Managing knowledge in organizations: An integrative framework and review of emerging themes, *Management Science* **49**(4): 571-582.

Armstrong, M. (2006). *A Handbook of Human Resource Management Practice*, Kogan Page, London and Philadelphia.

Armutat, S., Krause, H., Linde, F., Rump, J., Striening, W. & Weidmann, R. (2002). *Wissensmanagement erfolgreich einführen. Strategien – In-*

strument - Praxisbeispiele, Deutsche Gesellschaft für Personalführung e.V., Düsseldorf.

Autio, E., Sapienza, H. J., & Almeida, J. G. (2000). Effects of age at entry, knowledge intensity, and imitability on international growth, *Academy of Management Journal* **43**(5): 909-924.

Azevedo, A., Drost, E. A. & Mullen, M. R. (2002). Individualism and collectivism: Toward a strategy for testing measurement equivalence across culturally diverse groups, *Cross Cultural Management: An International Journal* **9**(1): 19-29.

Backhaus, K., Erichson, B., Plinke, W. & Weiber, R. (2011). *Multivariate Analysemethoden: Eine anwendungsorientierte Einführung*, Springer, Berlin/Heidelberg.

Badaracco, J. L., Jr. (1991). *The Knowledge Link. How Firms Compete Through Strategic Alliances*, Harvard Business School Press, Boston, Massachusetts.

Barley, S. R. (1983). Semiotics and the study of occupational and organizational cultures, *Administrative Science Quarterly* **28**(3): 393-413.

Barney, J. (1991). Firm resources and sustained competitive advantage, *Journal of Management* **17**(1): 99-120.

Baum, J. A. C. & Ingram, P. (1998). Survival-enhancing learning in the manhattan hotel industry, 1898-1980, *Management Science* **44**(7): 96-1016.

Becker, M. C. & Knudsen, M. P. (2006). Intra and inter-organizational knowledge transfer processes: Identifying the Missing Links, *DRUID Working Paper, No. 06-32*.

Benson, D. & Ziedonis, R. H. (2009). Corporate venture capital as a window on new technologies: Implications for the performance of corporate investors when acquiring startups, *Organization Science* **20**(2): 329-351.

Berk, R. A. (2004). *Regression Analysis: A Constructive Critique*, Sage Publications, Inc., Thousand Oaks.

Bettinger, C. (1989). Use corporate culture to trigger high performance, *Journal of Business Strategy* **10**(2): 38-42.

Bettis, R. A. & Prahalad, C. K. (1995). The dominant logic: Retrospective and extension, *Strategic Management Journal* **16**(1): 5-14.

Bhagat, R. S., Kedia, B. L., Harveston, P. D. & Triandis, H. C. (2002). Cultural variations in the cross-border transfer of organizational knowledge: An integrative framework, *Academy of Management Review* **27**(2): 204-221.

Brosius, F. (1998). *SPSS 8.0. Professionelle Statistik Unter Windows*, MITP-Verlag GmbH, Bonn.

Chatman, J. A. & Cha, S. E. (2003). Leading by leveraging culture, *California Management Review* **45**(4): 20-34.

Chatterjee, S., Lubatkin, M. H., Schweiger, D. M. & Weber, Y. (1992). Cultural differences and shareholder value in related mergers: Linking equity and human capital, *Strategic Management Journal* **12**(5): 319–334.

Chaudhuri, S. & Tabrizi, B. (1999). Capturing the real value in high-tech acquisitions, *Harvard Business Review* **77**(5):123-30.

Chen, C. C., Chen, X.-P. & Meindl, J. R. (1998). How can cooperation be fostered? The cultural effects of individualism-collectivism, *Academy of Management Review* **23**(2): 285-304.

Chesbrough, H. W. (2006). *Open Innovation: The New Imperative for Creating and Profiting from Technology*, Harvard Business School Publishing Corporation, Boston, Massachusetts.

Chesbrough, H., Vanhaverbeke, W. & West, J. (2006). *Open Innovation: Researching a New Paradigm*, Oxford University Press, Oxford, (U.K).

Choi, B. (2002). *Knowledge Management Enablers, Processes, and Organizational Performance: An Integration and Empirical Examination* (Thesis), Division of Management Engineering, Korea Advanced Institute of Science and Technology, Korea.

Churchill, G. A. (1979). A paradigm for developing better measures of marketing constructs. *Journal of Marketing Research*, **XVI**: 64–73.

Churchill, G. A. (1992). Better Measurement Practices Are Critical to Better Understanding of Sales Management Issues, *The Journal of Personal Selling and Sales Management* **12**(2): 73-80.

Clark, K. B. & Fujimoto, T. (1991). *Product Development Performance*, Harvard Business School Press, Boston.

Cohen, M. D., March, J. G. & Olsen, J. P. (1972). A garbage can model of organizational choice, *Administrative Science Quarterly* **17**: 1-25.

Cohen, W. M. & Levinthal, D. A. (1989). Innovation and learning: The Two faces of R & D, *The Economic Journal* **99**(397): 569-596.

Cohen, W. M. & Levinthal, D. A. (1990). Absorptive capacity: A new perspective on learning and innovation, *Administrative Science Quarterly* **35**(1): 128-152.

Cohen, W. M. & Levinthal, D. A. (1994). Fortune favors the prepared firm, *Management Science* **40**(2): 227-251.

Cohen, W. M. & Levinthal, D. A. (1997). Reply to "comments on 'fortune favors the prepared firm'", *Management Science* **43**(10): 1463-1468.

Cooke & Lafferty (1989). *Organizational Culture Inventory*, Human Synergistics International, Plymouth.

Costello, A. B. & Osborne, J. W. (2005). Best Practices in Exploratory Factor Analysis: Four Recommendations for Getting the Most From Your Analysis, *Practical Assessment Research & Evaluation* **10**(7): 1-9

Creswell, J. W. (2009). *Research Design: Qualitative, Quantitative, and Mixed Methods Approaches*, Sage Publication, London, U.K.

Cronbach, L. J. (1951). Coefficient alpha and the internal structure of tests, *Psychometrika* **16**(3): 297-334.

Crossan, M. M., Lane, H. W. & White, R. E. (1999). An organizational learning framework: From intuition to institution, *The Academy of Management Review* **24**(3): 522-537.

Crotty, M. (2003). *The Foundations of Social Research. Meaning and Perspective in the Research Process*, SAGE Publications LTD, London.

Culkin, N. & Smith, D. (2000). An emotional business: A guide to understanding the motivations of small business decision takers, *Qualitative Market Research: An International Journal* **3**(3): 145-157.

Darr, E. D., Argote, L. & Epple, D. (1995). The Acquisition, transfer, and depreciation of knowledge in service organizations: Productivity in franchises, *Management Science* **41**(11): 1750-1762.

Das, T. K. & Teng, B.-S. (1998). Between trust and control: Developing confidence in partner cooperation in alliances, *Academy of Management Review* **23**(3): 491-512.

Davenport, T. & Prusak, L. (1998). *Working Knowledge: How Organizations Manage What They Know*, Harvard Business School Press, Cambridge, MA.

Davenport, T. H., De Long, D. W. & Beer, M. C. (1997). Building Successful Knowledge Management Projects, *working Paper from the Center for Business Innovation^{SM}*, available at http://www.providersedge.com/Docs/Km_Articles/Building_Successfu l_KM_Projects.Pdf, last access 12/22/12.

Davenport, T. H., De Long, D. W. & Beers, M. C. (1998). Successful knowledge management projects, *Sloan Management Review* (Winter): 43-57.

Davis, F. B. (1964). *Educational Measurements and Their Interpretation*, Belmont, CA: Wadsworth.

Davis, L. E. (1966). The Design of Jobs, *Industrial Relations: A Journal of Economy and Society* 6(1): 21–45.

De Jong, J. P. J. & Marsili, O. (2005). The fruit flies of innovation: A taxonomy of innovative small firms, URL, http://www2.druid.dk/conferences/viewpaper.php?id=2756&cf=18, last access 09/16/12.

Deal, T. E. & Kennedy, A. A. (1982). *Corporate Cultures: The Rites and Rituals of Corporate Life*, Perseus Books Publishing, Jackson, U.S.

Del Carmen Haro-Domínguez, M., Arias-Aranda, D., Javier Lloréns-Montes, F., & Ruíz Moreno, A. (2007). The impact of absorptive capacity on technological acquisitions engineering consulting companies, *Technovation* **27**(8): 417–425.

Denison, D. R. & Mishra, A. K. (1995). Toward a theory of organizational culture and effectiveness, *Organization Science* **6**(2): 204-223.

Denison, D. R. (1990). *Corporate Culture and Organizational Effectiveness*, Wiley, New York, U.S.

Denning, S. (2005). Why the best and brightest approaches don't solve the innovation dilemma, *Strategy & Leadership* **33**(1): 4-11.

DeVellis, R. F. (2011). *Scale development: Theory and Applications (Applied Social Research Methods) (3rd ed.)*, Sage, Thousand Oaks.

DiBella, A. J. & Nevis, E. C. (1998). *How organizations learn. An integrated strategy for building learning capability*, Jossey-Bass Publishers, San Francisco.

Dijksterhuis, M. S., Van den Bosch, F. A. J. & Volberda, H. W. (1999). Where do new organizational forms come from? Management logics as a source of coevolution, *Organization Science* **10**(5): 569-582.

Ding, Q., Akoorie, M. & Pavlovich, K. (2009). A critical review of three theoretical approaches on knowledge transfer in cooperative alliances, *International Journal of Business and Management* **4**(1): 47-56.

Donaldson, L. (2001). *The Contingency Theory of Organizations*, Sage Publication s, Inc. Thousand Oaks.

Dyer, W. G. Jr., Dyer, J. H. & Dyer, W. G. (2003). *Team Building: Proven Strategies for Improving Team Performance*, John Wiley & Sons, San Francisco.

Dziuban, C. D. & Shirkey, E. C. (1974). When is a correlation matrix appropriate for factor analysis? some decision rules, *Psychological Bulletin* **81**(6) : 358-361.

Ebrahim, N. A., Ahmed, S. & Taha, Z. (2010). SMEs; Virtual research and development (R&D) teams and new product development: A literature review, International Journal of the Physical Sciences **5**(7): 916-930.

Edmondson, A. (1999). Psychological safety and learning behavior in work teams, *Administrative Science Quarterly*, **4**(2): 350-383.

Eisenhardt, K. M. & Jeffrey, A. M. (2000). Dynamic capabilities: what are they?, *Strategic Management* Journal **21**: 1105–1121.

European Commission (2012). Small and medium-sized enterprises (SMEs). What is an SME?, URL, http://ec.europa.eu/enterprise/policies/sme/facts-figures-analysis/sme-definition/index_en.htm, last access 10/08/12.

Evers, H.-D. (2002). Knowledge society and the knowledge gap, *Paper read at an International Conference, "Globalisation, Culture and Inequalities"*, Bangi, Malaysia.

Fahey, L. (1999). *Outwitting, Outmaneuvering, And Outperforming Competitors*, Wiley, New York.

Federal Office of Statistics (2008). *Klassifikation der Wirtschaftszweige Mit Erläuterungen*, Federal Office of Statistics, Wiesbaden.

Feinberg, S. E. & Gupta, A. L. (2004). Knowledge spillovers and the assignment of R&D responsibilities to foreign subsidiaries, *Strategic Management Journal* **25**: 823-845.

Field, A. (2013). *Discovering Statistics Using IBM SPSS Statistics*, SAGE Publications Ltd., Los Angeles/London/New Delhi/Singapore/Washington DC.

Fiol, C. M. & Lyles, M. A. (1985). Organizational learning, *The Academy of Management Review* **10**(4): 803-813.

Flatten, T. C., Engelen, A., Zahra, S. A. & Brettel, M. (2011a). A measure of absorptive capacity: Scale development and validation, *European Management Journal* **29**: 98-116.

Flatten, T. C., Greve, G. I. & Brettel, M. (2011b). Absorptive Capacity and Firm Performance in SMEs: The Mediating Influence of Strategic Alliances, *European Management Review* **8**(3): 137-152.

Fletcher, B. & Jones, F. (1992). Measuring organizational culture: The cultural audit, *Managerial Auditing Journal* **7**(6): 30-36.

Ford, D. & Chan, Y. (2002). Knowledge sharing in a cross-cultural setting: A case study, *Queen's University at Kingston Working Paper, WP 02-09*.

Fox, A (1973). *Beyond Contract*, Faber and Faber, London.

Frey, D. & Schulz-Hardt, S. (2000). *Vom Vorschlagswesen zum Ideenmanagement: Zum Problem der Änderungen von Mentalitäten, Verhalten und Strukturen*, Verlag für Angewandte Psychologie, Göttingen.

Gambetta, D. (1988). *Trust: Making and Breaking Cooperative Relations*, Blackwell Publishing, New York.

Garvin, D. A. (1993). Building a learning organization, *Harvard Business Review* **73**(4): 78-91.

Gaylin, W. (1976). *Caring*, Alfred Knopf, New York.

Girdauskienė, L. & Savanevičienė, A. (2007). Influence of knowledge culture on effective knowledge transfer, *ISSN 1392-2785 ENGINEERING ECONOMICS. 2007*, **4**(54): 36-43.

Glaser, S. R. (1983). *Assessing organizational culture: An interpretive approach*, Paper presented at the Speech Communication Association Convention, Washington, DC.

Glisby, M. & Holden, N. (2003). Contextual constraints in knowledge management theory: the cultural embeddedness of Nonaka's knowledge-creating company, *Knowledge and Process Management* **10**(1): 29-36.

Glückstein, S. (2000). *Wissensmanagement - Eine neo-institutionalistische Perspektive*, GRIN Verlag, München.

Gordon, G. G. & Cummins, W. M. (1979). *Managing Management Climate*, Lexington Books, Lexington, Mass.

Gordon, G. G. & DiTomaso, N. (1992). Predicting corporate performance from organizational culture, *Journal of Management Studies* **29**(6): 783-798.

Granstrand, O. & Sjölander, S. (1990). The acquisition of technology and small firms by large firms, *Journal of Economic Behavior & Organization* **13**(3): 367-386.

Grant, R. M. (1996a). Toward a knowledge-based theory of the firm, *Strategic Management Journal* **17**(Winter Special Issue): 109-122.

Grant, R. M. (1996b). Prospering in dynamically-competitive environments: Organizational capability as knowledge integration, *Organization Science* **7**(4): 375-387.

Gruenfeld, D. H., Martorana, P. V. & Fan, E. T. (2000). What do groups learn from their worldliest members? Direct and indirect influence in dynamic teams, *Organizational Behavior and Human Decision Processes* **82**(1): 45–59.

Hackman, J. R. & Oldham, G. R. (1976). Motivation through the design of work: Test of a theory, *Organizational Behavior and Human Performance* **16**: 250-279.

Hall, R. (2003). *Knowledge Management in the New Business Environment*, Acirrt, University of Sydney, Sydney.

Harrison, R (1972). Understanding your organisation's culture, *Harvard Business Review* **50**(3): 119-128.

Hatten, K. & Rosenthal, S. (2002). Knowledge management: creating a knowing culture, URL, http://iveybusinessjournal.com/topics/governance/knowledge-management-creating-a-knowing-culture#.UoRuZeKmYlQ, last access 14/02/13.

Henderson, J. C., & Lee, S. (1992). Managing I/S design teams: A control theories perspective, *Management Science* **38**(6): 757-777.

Hofstede, G. (1980). *Culture's Consequences: International Differences in Work-Related Values*, SAGE Publicationsc Beverly Hills.

Hofstede, G. (1998). Attitudes, values and organizational culture: Disentangling the concepts, *Organization Studies* **19**(3): 477-493.

Hofstede, G., Neuijen, B., Ohayv, D. D. & Sanders, G. (1990). Measuring organizational cultures: A qualitative and quantitative study across twenty cases, *Administrative Science Quarterly* **35**(2): 286-316.

Holden, N. J. & von Kortzfleisch, H. F. O. (2004). Why cross-cultural knowledge transfer is a form of translation in more ways than you think, *Knowledge and Process Management* **11**(2): 127-138.

Holden, N. J. (2001). Knowledge management: Raising the spectre of the cross-cultural dimension, *Knowledge and Process Management* **8**(3): 155-163.

Holmes, S. & Marsden, S. (1996). An exploration of the espoused organisational cultures of public accounting firms, *Accounting Horizons* **26**(10): 26-35.

Huemer, L., Krogh. G. & Johan, R. (1998). Knowledge and the concept of trust, *in* Krogh, G., Roos, J. & Kleine, D., *Knowing in Firms*, Sage, Thousand Oaks: 123-145.

Hurley, R. F. & Hult, T. M. (1998). Innovation, market orientation, and organizational learning: an integration and empirical examination, *Journal of Marketing* **62**: 42-54.

Huygens, M., Baden-Fuller, C., Van den Bosch, F. A. J. & Volberda, H. W. (2001). Co-evolution of firm capabilities and industry competition: Investigating the music industry, 1877-1997, *Organization Studies* **22**(6): 971-1011.

ILOI (1997). *Knowledge Management: ein empirisch gestützter Leitfaden zum Management des Produktionsfaktors Wissen*, ILOI,München.

Ingram, P. & Baum, J. A. C. (1997). Chain affiliation and the failure of Manhattan hotels, 1898-1980, *Administrative Science Quarterly* **42**(1): 68-102.

Institute for SME Research Bonn (2007). Der Mittelstand in der Bundesrepublik Deutschland: Eine volkswirtschaftliche Bestandsaufnahme, URL, http://www.ifm-bonn.org/assets/documents/BMWI-Dokumentation-561.pdf, last access 02/24/13.

Institute for SME Research Bonn (2012). KMU-Definition des IfM Bonn, URL, http://www.ifm-bonn.org/index.php?id=89, last access 10/08/12.

Ivens, S., Schaarschmidt, M. & Zerwas, D. (2014). Absorptive capacity and the complementarity of control mechanisms in open source software development: A knowledge-based-view (Research in Progress), *Poster for the MKWI 2014* (forthcoming) and *Proceedings of the MKWI 2014* (forthcoming), Paderborn.

Janićijević, N. (2011). Methodological Approaches in the Research of Organizational Culture, Economic Annals **LVI**(189): 69-99.

Jansen, J. J. P., Van den Bosch, F. S. J. & Volberda, H. W. (2005). Managing potential and realized absorptive capacity: How do organizational antecedents matter?, *Academy of Management Journal*, **48**(6): 999-1015.

Janz, B. D. & Prasarnphanich, P. (2003). Understanding the Antecedents of Effective Knowledge Management: The Importance of a Knowledge-Centered Culture, *Decision Sciences* **34**(2): 351-384.

Jolliffe, I. T. (2002). *Principal Component Analysis*, Springer-Verlag, New York.

Jones, G. R. & George, J. M. (1998). The experience and evolution of trust: Implications for cooperation and teamwork, *Academy of Management Review* **23**(3): 531-546.

Kaiser, H. F. (1970). A *second generation little jiffy*, Psychometrika **35**(4):401-415.

Kaiser, H. F. (1974). An index of factorial simplicity, *Psychometrika* **39**(1): 31-36.

Kaplan, D. (2000). *Structural Equation Modeling. Foundations and Extensions*, Sage Publications, Inc., Thousand Oaks.

Kaplan, R. W. & Saccuzzo, D. P. (1982). Psyc*hological Testing: Principles, Applications and Issues, Monterey*, CA, Brooks/Cole.

Katz, R. & Allen, T. J. (1982). Investigating the Not Invented Here (NIH) syndrome: A look at the performance, tenure, and communication patterns of 50 R & D Project Groups, *R&D Management* **12**(1): 7-20.

Katz, R. & Tushman, M. (1981). An investigation into the managerial roles and career paths of gatekeepers and project supervisors in a major R & D facility, *R & D Management* **11**(3): 103-110.

Katzy, B. & Klein, S. (2008). Editorial introduction - Special issue on living labs, *The Electronic Journal for Virtual Organizations and Networks* **10**: 2-6.

Kedia, B. L. & Bhagat, R. S. (1988). Cultural constraints on transfer of technology across nation, *The Academy of Management Review* **13**(4): 559-571.

Kern, H. (1991). *Analyse von Unternehmenskulturen. Eine empirische Studie*, Peter Lang, Frankfurt am Main.

Kieser, A. & Ebers, M. (2006). *Organisationstheorien*, Kohlhammer GmbH, Stuttgart.

Kilian, T. (2005). *Kundenorientierung von Energieversorgungsunternehmen: Konzepte, Messung und Umsetzung*, Verlag Dr. Köster, Berlin.

Kilmann, R. H. & Saxton M. J. (1983). *The Kilmann-Saxton Culture-Gap Survey*, Organizational Design Consultants, Pittsburgh.

Kim, L. (1998). Crisis construction and organizational learning: Capability building in catching-up at Hyundai Motor, *Organization Science* **9**(4): 506-521.

King, W. R. (2007). A research agenda for the relationships between culture and knowledge management, *Knowledge and Process Management* **14**(3): 226-236.

Kleinbaum, D. G., Kupper, L. L. & Muller, K. E. (2008). *Applied Regression Analysis and Other Multivariable Methods (Duxbury Applied Series)*, Duxbury (Thomson Higher Education), Belmont.

Knudsen, M. P., Dalum, B. & Villumsen, G. (2001). Two faces of absorptive capacity creation: Access and utilisation of knowledge, Preliminary Draft! To be presented at the Nelson and Winter Conference organised by DRUID in Aalborg, Denmark, June 2001, URL, http://www.google.de/url?sa=t&rct=j&q=&esrc=s&source=web&cd=1 &cad=rja&ved=0CDMQFjAA&url=http%3A%2F%2Fciteseerx.ist.psu .edu%2Fviewdoc%2Fdownload%3Fdoi%3D10.1.1.195.3077%26rep%

3Drep1%26type%3Dpdf&ei=nxENUa_UDtDJsga3mYHwBw&usg=A
FQjCNGnKRi1semlnBvg8odTt6dHAdGF-
w&bvm=bv.41867550,d.Yms, last access 02/02/13.

Koberg, C. S. & Chusmir, L. H. (1987). Organizational culture relationships with creativity and other job-related variables, *Journal of Business Research* **15**: 397-409.

Kobi, J.-M. & Wüthrich, H. A. (1986). *Unternehmenskultur verstehen, erfassen und gestalten*, verlag moderne industrie AG & Co. Buchverlag, Landesberg/Lech.

Koestler, A. (1964). *The Act of Creation*, Hutchinson, London, U.K.

Kogut, B. & Zander, U. (1992). Knowledge of the firm, combinative capabilities, and the replication of technology, *Organization Science* **3**(3): 383-397.

Kogut, B. & Zander, U. (1993). Knowledge of the firm and the evolutionary theory of the multinational corporation, *Journal of International Business Studies* **24**: 625-645.

Kotter, J. P. & Heskett, J. L. (1992). *Corporate Culture and Performance*, The Free Press, New York, U.S.

Kumar, N., Stern, L. W. & Anderson, J. C. (1993). Conducting interorganizational research using key informants, *The Academy of Management Journal* **36**(6): 1633-1651.

Kunz, J. (2010). Gestaltungsparameter zur Förderung des innerbetrieblichen Wissenstransfers – Ein Strukturierungsrahmen, *Zeitschrift für Management* **5**: 29-52.

Lane, P. J. & Lubatkin, M. (1998). Relative absorptive capacity and interorganizational learning, *Strategic Management Journal* **19**(5): 461-477.

Lane, P. J., Koka, B. R. & Pathak, S. (2006). The reification of absorptive capacity: A critical review and rejuvenation of the construct, *The Academy of Management Review* **31**(4): 833-863.

Lane, P. J., Salk, J. E & Lyles, M. A. (2001). Absorptive capacity, learning, and performance in international joint ventures, *Strategic Management Journal* **22**: 1139–1161.

Larsson, R., Bengtsson, L., Henriksson, K. & Sparks, J. (1998). The interorganizational learning dilemma: Collective knowledge development in strategic alliances, *Organization Science* **9**(3): 285-305.

Laursen, K. & Salter, A. (2006). Open for innovation: The role of openness in explaining innovation performance among U.K. manufacturing firms, *Strategic Management Journal* **27**: 131-150.

Lawrence, J. E. (2008). The Challenges and utilization of e-commerce: The use of internet by small to medium-sized enterprises in the United Kingdom, *Information, Society and Justice* **1**(2): 99-113.

Lawrence, J. E. (2010). The strategic business value of internet usage in small to medium-sized enterprises, Information, *Society and Justice* **3**(1): 37-50.

Lee, H. & Choi, B. (2003). Knowledge management enablers, processes, and organizational performance. An integrative view and empirical examination, *Journal of Management Information Systems* **20**(1): 179-228.

Lemken, B., Kahler H. & Rittenbruch M. (2000). Sustained knowledge management by organizational culture, *Proceedings of the Hawai'I International Conference On System Sciences*, Maui, Hawaii.

Lenox, M. & King, A. (2004). Prospects for developing absorptive capacity through internal information provision, *Strategic Management Journal* **25**: 331-345.

Leonard-Barton, D. (1995). *Wellsprings of knowledge*, Harward Business School Press, Boston.

Leroy, G. (2011). *Designing User Studies in Informatics*, Springer-Verlag, London.

Levine, J. M., Higgins, E. T. & Choi, H.-S. (2000). Development of strategic norms in groups, *Organizational Behavior and Human Decision Processes* **82**(1): 88-10.

Levitt, B. & March, J. G. (1988). Organizational Learning, *Annual Review of Sociology* **14**: 319-340.

Lewin, A. Y. & Volberda, H. W. (1999). Prolegomena on coevolution: A framework for research on strategy and new organizational forms, *Organization Science* **10**(5): 519-534.

Lewin, A. Y., Long, C. P. & Carroll, T. N. (1999). The coevolution of new organizational forms, *Organization Science* **10**(5): 535-550.

Lewin, A. Y., Massini, S. & Peeters, C. (2011). Microfoundations of internal and external absorptive capacity routines, *Organization Science* **22**(1): 81-98.

Liao, J., Welsch, H. & Stoica, M. (2003). Organizational absorptive capacity and responsiveness: An empirical investigation of growth-oriented SMEs, *Entrepreneurship Theory and Practice* **28**(1): 63-85.

Liao, S.-h., Fei, W.-C. & Chen, C.-C. (2007). Knowledge sharing, absorptive capacity, and innovation capability: An empirical study of Taiwaan's knowledge-intensive industries, *Journal of Information Science* **33**(3): 340-359.

Lichtenthaler, U. & Lichtenthaler, E. (2009). A Capability-Based Framework for Open Innovation: Complementing Absorptive Capacity, *Journal of Management Studies* **46**(8): 1315-1338.

Lichtenthaler, U. & Lichtenthaler, E. (2012). Technology transfer across organizational boundaries: Absorptive capacity and desorptive capacity, *California Management Review* **53**(1): 154-170.

Lichtenthaler, U. (2009). Absorptive capacity, environmental turbulence, and the complementarity of organizational learning processes, *Academy of Management Journal* **52**(4): 822-846.

Liebowitz, J. (1999). Key ingredients to the success of an organization's knowledge management strategy, *Knowledge and Process Management* **6**(1): 37-40.

Lim, B. (1995). Examining the organizational culture and organizational performance link, *Leadership & Organization Development Journal* **16**(5): 16-21.

Lucas, L. M. (2006). The role of culture on knowledge transfer: The case of the multinational corporation, *The Learning Organization* **13**(3): 257-275.

Lyles, M. A. & Salk, J. E. (1996). Knowledge acquisition from foreign partners in international joint ventures, *Journal of International Business Studies* **27**(5): 877-904.

Lyles, M. A. & Schwenk, C. R. (1992). Top management, strategy and organizational knowledge structures, *Journal of Management Studies* **29**(2): 155-174.

Maddux, R. & Wingfield, B. (2003). *Team Building: An Exercise in Leadership*, Crisp Publications, Inc., Seattle, Washington, United States.

Malhotra, A., Gosain, S. & El Sawy, O. A. (2005). Absorptive capacity configurations in supply chains: Gearing for partner-enabled market knowledge creation, *MIS Quarterly* **29**(1): 145-187.

Manz, C. C. & Sims, Jr., H. P. (1980). Self-management as a substitute for leadership: A social learning theory perspective, *The Academy of Management Review* **5**(3): 361-367.

Manz, C. C. (1992). Self-leading work teams: Moving beyond self-management myths, *Human Relations* **45**(11): 1119-1140.

Mason, R. D. & Mitroff, I. (1981). *Challenging Strategic Planning Assumptions*, Wiley, New York.

Massa, S. & Testa, S. (2008). Innovation and SMEs: Misaligned perspectives and goals among entrepreneurs, academics, and policy makers, *Technovation* **28**: 393-407.

Matijevic, A. (2005). *Defining the Web-Business am Beispiel deutscher Energieversorgungsunternehmen. Status quo und Entwicklungsperspektiven, strategische Stoßrichtungen und empirische Fundierung*, vme Verlag und Medienservice Energie Jürgen Pöschk.

Matusik, S. F. & Heeley, M. B. (2005). Absorptive capacity in the software industry: Identifying dimensions that affect knowledge and knowledge creation activities, *Journal of Management* **31**(4): 549-572.

Maurer, C. & Tiwana, A. (2012). Control in app platforms: The integration-differentiation paradox, *Thirty Third International Conference on Information Systems*, Orlando 2012.

Mayer, R. C., Davis, J. H. & Schoorman, F. D. (1995). An Integrative Model of Organizational Trust, *The Academy of Management Review* **20**(3): 709-734.

Mayeroff, M. (1971). *On Caring*, Harper and Row, New York.

Maynard, M. (2002). *Researching Women's Lives From A Feminist Perspective*, Taylor & Francis, Oxfordshire.

McCall, M. W. & Kaplan, R. E. (1985). *Whatever It Takes: Decision Makers at Work*, Prentice-Hall, Inc., Englewood Cliffs, New Jersey.

McEvily, B. & Zaheer, A. (1999). Bridging ties: a source of firm heterogeneity in competitive capabilities, *Strategic Management Journal* **20**(12): 1133-1156.

McEvily, B., Perrone, V. and Zaheer, A. (2003). Trust as an organizing principle, *Organization Science* **14**(1): 91-103.

McGrath, R. G. & MacMillan, I.C. (2000). *The Entrepreneurial Mindset: Strategies for Continuously Creating Opportunity in an Age of Uncertainty*, Harvard Business School Press, Boson, Mass.

McKnight, P. E., McKnight, K. M., Sidani, S. & Figueredo, A. J. (2007). *Missing Data: A Gentle Introduction (Methodology in the Social Sciences)*, The Guilford Press, New York.

Meeus, M. T. H., Oerlemans, L. A. G., & Hage, J. (2001). Patterns of interactive learning in a high-tech region, *Organization Studies* **22**: 145-172.

Meyerson, D., Weick, K. E. & Kramer, R. M. (1996). Swift trust and temporary groups, *in* Kramer, R. M. & Tyler, T. R., *Trust in Organizations: Frontiers of Theory and Research*, Sage Publications, Thousand Oaks: 166-195.

Miles, G., Miles, R. E., Perrone, V. & Edvinsson, L. (1998). Some conceptual and research barriers to the utilization of knowledge, *California Management Review* **40**(3): 281-288.

Miles, R. E. & Snow, C. C. (1992). Causes of failure in network organizations, *California Management Review*: 53-72.

Miller, D. (1996). A preliminary typology of organizational learning: synthesizing the literature, *Journal of Management* **22**(3): 485-505.

Minbaeva, D. B. (2007). Knowledge transfer in multinational corporations, *Management International Review* **47**(4): 567-593.

Minbaeva, D., Pedersen, T., Björkman, I., Fey, C. F. & Park, H. J. (2003). MNC knowledge transfer, subsidiary absorptive capacity, and HRM, *Journal of International Business Studies* **34**: 586–599.

Mintzberg, H., Raisinghani, D. & Théorêt, A. (1976). The structure of 'unstructured' decision processes, *Administrative Science Quarterly* **21**(2): 246-275.

Moffett, S., McAdam, R. & Parkinson, S. (2002). Developing a model for technology and cultural factors in knowledge management: a factor analysis, *Knowledge and Process Management* **9**(4): 237-255.

Molina, L. M. & Lloréns-Montes, F. J. (2006). Autonomy and teamwork effect on knowledge transfer – knowledge transferability as a moderator variable, *International Journal of Technology Transfer and Commercialization* **3**: 264-280.

Moreland, R. L. & Myaskovsky, L. (2000). Exploring the performance benefits of group training: Transactive memory or improved communication? *Organizational Behavior and Human Decision Processes* **82**(1): 117-133.

Morris, M. H., Kuratko, D. F. & Covin, J. G. (2008). *Corporate Entrepreneurship & Innovation*, Florence: Thomson South-Western, Mason, USA.

Mowery, D. C., Oxley, J. E. (1995). Inward technology transfer and competitiveness: The role of national innovation systems, *Cambridge Journal of Economics* **19**(1): 67-93.

Mowery, D. C., Oxley, J. E., & Silverman, B. S. (1996). Strategic alliances and interfirm knowledge transfer, *Strategic Management Journal* **17**: 77-91.

Murphy, K. R. & Davidshofer, C. O. (1988). *Psychological Testing: Principles and Application*, Englewood Cliffs, NJ, Prentice-Hall.

Nahapiet, J. & Ghoshal, S. (1998). Social capital, intellectual capital, and the organizational advantage, *The Academy of Management Review* **23**(2): 242-266.

Ndiege, J. R., Herselman, M. E. & Flowerday, S. V. (2012). Absorptive capacity: Relevancy for large and small enterprises, *SA Journal of Information Management* **14**(1): 1-9.

Nelson, K. M. & Cooprider, J. G. (1996). The contribution of shared knowledge to IS group performance, *MIS Quarterly*: 409-432.

Nelson, R. R. & Winter, S. G. (1982). *An Evolutionary Theory of Economic Change*, Harvard University Press, Cambridge, MA.

Newell, S., Robertson, M., Scarbrough, H. & Swan, J. (2009). *Managing Knowledge Work and Innovation. Second Edition*, Basingstoke, Palgrave Macmillan.

Nonaka, I. & Takeuchi, H. (1995). *The Knowledge Creating Company*, Oxford University Press, New York.

Nonaka, I. (1994). A dynamic theory of organizational knowledge creation, *Organization Science* **5**(1): 14–37.

Nonaka, I. (2007). The Knowledge-Creating Company, *Harvard Business Review* (July–August): 162-171.

Nonaka, I., Umemoto, K. & Senoo, D. (1996). From information processing to knowledge creation: A paradigm shift in business management, *Technology in Society* **18**(2): 203-218.

Nunnally, J. C. (1967). *Psychometric Theory*, 1st ed., New York, McGraw-Hill.

Nunnally, J. C. (1978). *Psychometric Theory*, 2d ed., New York, McGraw-Hill.

O'Reilly, C. A. & Roberts, A. H. (1974). Information filtration in organizations: Three experiments, *Organizational Behavior and Human Performance* **11**: 253-265.

O'Reilly, C., Chatman, J. & Caldwell, D. (1991). People and organizational culture: A profile comparison approach to assessing person-organization fit, *Academy of Management Journal* **34**(3): 487-516.

O'Dell, C. & Grayson Jr., C. J. (1999). Knowledge transfer: Discover your value proposition, *Strategy & Leadership* **27**(2): 10-15.

OECD Centre for Entrepreneurship, SMEs and Local Development (2009). *The Impact of the Global Crisis on SME and Entrepreneurship Financing and Policy Responses: Contribution to the OECD Strategic Response to the Financial and Economic Crisis*, BEL CANTO, Boulogne, France.

Orlikowski, W. J. (2002). Knowing in practice: Enacting a collective capability in distributed organizing, *Organization Science* **13**(3): 249-273.

Parker, G. M. & Kropp, R. P. (1992). *50 Activities for Team Building*, HRD Press, Amherst, Massachusetts, United States.

Pemberton, J. D. & Stonehouse, G. H. (2002). The importance of individual knowledge in developing the knowledge-centric organization, *in*

Coakes, E., Willis, D. & Clarke, S. (Eds), *Knowledge Management in the SocioTechnical World. The Graffiti Continues*, Springer-Verlag London Limited 2002, London: 77-89.

Penrose, E. T. (1959). *The Theory of the Growth of the Firm*, New York, NY, Oxford University Press, New York.

Peter, J. P. (1979). Reliability: A Review of Psychometric Basics and Recent Marketing Practices, *Journal of Marketing Research* 16(1): 6-17.

Peters, T. & Waterman, R. (1982). *In Search of Excellence*, Harper & Row, New York, U.S.

Peterson, R. A. (1994). A Meta-Analysis of Cronbach's Coefficient Alpha, *Journal of Consumer Research* 21(2): 381-391.

Picot, A., Reichwald, R. & Wigand, R. T. (2003). *Die grenzenlose Unternehmung: Information, Organisation und Management: Lehrbuch zur Unternehmensführung im Informationszeitalter*, Gabler, Wiesbaden, Germany.

Poech, A. (2003). *Erfolgsfaktor Unternehmenskultur : eine empirische Analyse zur Diagnose kultureller Einflussfaktoren auf betriebliche Prozesse*, Utz, München.

Powell, W. W., Koput, K. W. & Smith-Doerr, L. (1996). Interorganizational collaboration and the locus of innovation: Networks of learning in biotechnology, *Administrative Science Quarterly* 41: 116-145.

Prahalad, C. K. & Bettis, R. A. (1986). The dominant logic: A new linkage between diversity and performance, *Strategic Management Journal* 7(6): 485-501.

Probst, G. J. B. (1998). Practical Knowledge Management: A Model That Works, *Prism, Arthur D. Little* (Second Quarter): 17-29.

Probst, G., Raub, S. & Romhardt, K. (2010). *Wissen Management. Wie Unternehmen ihre wertvollste Ressource optimal nutzen (6. Auflage)*, Gabler |GWV Fachverlage GmbH, Wiesbaden.

Quick, T. L. (1992). *Successful Team Building*, AMACOM, a division of American Management Association, New York.

Reagans, R. & McEvily, B. (2003). Network structure and knowledge transfer: The effects of cohesion and range, *Administrative Science Quarterly* **48**(2): 240-267.

Reynolds, P. (1986). Organizational culture as related to industry, position and performance: A preliminary report, *Journal of Management Studies* **23**(3): 333-345.

Rigby, D. & Zook, C. (2002). Open-market innovation, *Harvard Business Review*: 80-89.

Robbins, R. F. (2003). Harnessing "group memory" to build a knowledge-sharing culture, URL, http://www.accessmylibrary.com/article-1G1-103448677/harnessing-group-memory-build.html, last access 01/28/13.

Roberts, N., Galluch, P. S., Dinger, M., and Grover, V. (2012). Absorptive capacity and information systems research: review, synthesis, and directions for future research, *MIS Quarterly* **36**(2): 625-648.

Rocha, F. (1999). Inter-firm technological cooperation – Effects of absorptive capacity, firm-size and specialization, *Economics of Innovation and New Technology* **8**: 253-271.

Romano, J. P., Shaikh, A. M. & Wolf, M. (2010). *Multiple Testing*, URL, http://home.uchicago.edu/amshaikh/webfiles/palgrave.pdf, last access 10/19/13.

Rosenkopf, L. & Nerkar, A. (2001). Beyond local search: Boundary-spanning, exploration, and impact in the optical disk industry, *Strategic Management Journal* **22**(4): 287-306.

Roskin, R. (1986). Corporate culture revolution: the management development imperative, *Journal of Managerial Psychology* **1**(2): 3-9.

Rothaermel, F. T. & Alexandre, M. T. (2009). Ambidexterity in technology sourcing: The moderating role of absorptive capacity, *Organization Science* **20**(4): 759-780.

Rousseau, D. M., Sitkin, S. B., Burt, R. S. & Camerer, C. (1998). Not so different after all: A cross-discipline view of trust, *Academy of Management Review* **23**(3) 393-404.

Rowley, J. (2007). *The wisdom hierarchy: representations of the DIKW hierarchy*, Journal of Information Science **33**(2): 163-180.

Sashkin, M. (1984). *Organizational Beliefs Questionnaire: Pillars of Excellence*, Bryn Mawr, Pennsylvania.

Schein, E. H. (1981). Does Japanese management style have a message for American managers?, *Sloan Management Review* **23**(1): 55-68.

Schein, E. H. (1983). The role of the founder in creating organizational culture, *Organizational Dynamics*: 13-28.

Schein, E. H. (1984). Coming to a new awareness of organizational culture, *Sloan Management Review* **25**(2): 3-16.

Schein, E. H. (1985). *Organizational Culture and Leadership*, Jossey-Bass, San Francisco.

Schein, E. H. (1990). Organizational Culture, *American Psychologist* **45**(2): 109-119.

Scholl, W., König, C., Meyer, B. & Heisig, P. (2004). The future of knowledge management. An international Delphi study, *Journal of Knowledge Management* **8**(2): 19-35.

Schwenk, C. R. (1984). The cognitive perspective on strategic decision making, *Journal of Management Studies* **25**(1): 41-56.

Seidler, J. (1974). On Using Informants: A Technique for Collecting Quantitative Data and Controlling Measurement Error in Organization Analysis, *American Sociological Review* **39**(6): 816-831.

Shane, S. (2000). Prior knowledge and the discovery of entrepreneurial opportunities, *Organization Science* **11**(4): 448–469.

Simonin, B. L. (1999). Ambiguity and the process of knowledge transfer in strategic alliances, *Strategic Management Journal* **20**(7): 595-623.

Smircich, L. (1983). Concepts of culture and organizational analysis, *Administrative Science Quarterly* **28**(3): 339-358.

Smith, K. G. & Di Gregorio, D. (2006). Bisociation, discovery, and the role of entrepreneurial action, *in* Hitt, M. A., Ireland, R. D., Camp, S. M. & Sexton, D. L., *Strategic Entrepreneurship. Creating a New Mindset*, Blackwell Publishing, Malden: 129-150.

Sollberger, B. A. (2002), *Wissenskultur: Erfolgsfaktor für ein ganzheitliches Wissensmanagement*, Haupt Verlag, Bern.

Spender, J.-C. (1996). Making knowledge the basis of a dynamic theory of the firm, *Strategic Management Journal* **17**: 45-62.

Spieth, P. (2009). *Wissenstransfer unternehmenskulturinduzierter Akteure: Eine multidimensionale Analyse der Unternehmenskultur als Einflussfaktor für den erfolgreichen Transfer von Wissen in Unternehmen*, Cactus Group Verlag, Kassel.

Starbuck, W. H. & Milliken, F. J. (1988). Executive's perceptual filters: What they notice and how they make sense, *in* Hambrick, D. (ed.), T*he Executive Effect: Concepts and Methods for Studying Top Managers*, JAI Press, Greenwich, CT: 35–65.

Stasser, G. Vaughan, S. I. & Stewart, D. D. (2000). Pooling unshared information: The benefits of knowing how access to information is distributed among group members, *Organizational Behavior and Human Decision Processes* **82**(1): 102-116.

Stata, R. (1994) Organizational learning – The key to management innovation, *in* Schneier, C. E., Russel, C. J. & Beatty, R. W., *The Training and Development Sourcebook*: 31-42.

Stewart, D. W. (1981). The application and misapplication of factor analysis in marketing research, *Journal of Marketing Research* **18**(1): 51-62.

Stier, W. (1999). *Empirische Forschungsmethoden (2. Auflage)*, Springer-Verlag Berlin Heidelberg, Berlin Heidelberg.

Szulanski, G. (1996). Exploring internal stickiness: Impediments to the transfer of best practice within the firm, *Strategic Management Journal* **17**: 27-43.

Tang, F., Xi, Y. & Ma, J. (2006). Estimating the effect of organizational structure on knowledge transfer: A neural network approach, *Expert Systems with Applications* **30**: 796-800.

Tang, S. (2005). Knowledge as Production factor: Toward a unified theory of economic growth, URL, http://yataisuo.cass.cn/UploadFile/2005102203439560.pdf, last access 09/09/12.

Teece, D. J. (1981). The multinational enterprise: Marketing failure and market power considerations, *Sloan Management Review* **22**(3): 3-7.

Teece, D. J., Pisano, G. & Shuen, A. (1997). Dynamic Capabilities and Strategic Management, *Strategic Management Journal* **18**(7): 509-533.

Thompson, L., Gentner, D. & Loewenstein J. (2000). Avoiding missed opportunities in managerial life: Analogical training more powerful than individual case training, *Organizational Behavior and Human Decision Processes* **82**(1): 60-75.

Todorova, G. & Durisin, B. (2007). Absorptive Capacity: Valuing a Reconceptualization, *Academy of Management Review* **32**(3): 774-786.

Topchick, G. S. (2001). *Managing Workplace Negativity*, AMACOM, New York, United States of Amercia.

Tsai, W. (2001). Knowledge Transfer in intraorganizational networks: Effects of network position and absorptive capacity on business unit innovation and performance, *Academy of Management Journal* **44**(5): 996-1004.

Unterreitmeier, A. (2004). *Unternehmenskultur bei Mergers & Acquisitions : Ansätze zu Konzeptualisierung und Operationalisierung*, Deutscher Universität-Verlag/GWV Fachverlage GmbH, Wiesbaden.

Uzzi, B. (1997). Social structure and competition in interfirm networks: the paradox of embeddedness, *Administrative Science Quarterly* **42**(1): 35-67.

Van de Vrande, V., de Jong J.P.J., Vanhaverbeke, W. & de Rochemont, M. (2008). *Open innovation in SMEs: Trends, motives and management challenges*, SCALES-initiative (SCientific AnaLysis of Entrepreneurship and SMEs), Zoetermeer, Netherlands.

Van den Bosch, F. A. J., Volberda, H. W. & De Boer, M. (1999). Coevolution of firm absorptive capacity and knowledge environment: Organizational forms and combinative capabilities, *Organization Science* **10**(5): 551-568.

Vanhaverbeke, W. (2012). *OPEN INNOVATION IN SMEs: How can small companies and start-ups benefit from open innovation strategies? (Research Report)*, Vlerick Leuven Ghent Management School, Leuven, Belgium.

Verba, S., Nie, N. H.& Kim, J.-o. (1980). *Participation and Political Equality: A Seven-Nation Comparison*, Cambridge University Press, Cambridge.

Vermeulen, F. & Barkema, H.G. (2001). Learning through acquisitions, *Academy of Management Journal* **44**: 457-476.

Volberda, H. W. & Lewin, A. Y. (2003). Co-evolutionary dynamics within and between firms: From evolution to co-evolution, *Journal of Management Studies* **40**(8): 2111-2136.

Volberda, H. W., Foss, N. J. & Lyles, M. A. (2009). Absorbing the concept of absorptive capacity: How to realize its potential in the organization field, *SMG (Center for strategic Management and Globalization) Working Paper*, No. 10/2009.

Volberda, H. W., Foss, N. J. & Lyles, M. A. (2010). Absorbing the concept of absorptive capacity: How to realize its potential in the organization field, *Organization Science* **21**(4): 931-951.

Von Hippel, E. (1988). *Sources of Innovation*, MIT Press, Cambrigde, MA.

Von Kortzfleisch, H. F. O. (2004). *Organisatorisch Balancierung von Informations- und Kommunikationstechnologien*, JOSEF EUL Verlag GmbH, Lohmar.

Von Krogh, G. (1998). Care in Knowledge Creation, *California Management Review* **40**(3): 133-153.

Walsh, J. P. (1995). Managerial and organizational cognition: notes from a trip down memory lane, *Organization Science* **6**(3): 280-321.

Wanberg, C. (2012). *The Oxford Handbook of Organizational Socialization*, Oxford University Press, Inc., New York.

Wang, C. L. & Ahmed, P. K. (2007). Dynamic capabilities: A review and research agenda, *International Journal of Management Reviews* **9**(1): 31-51.

Warkentin, M. & Beranek, P. M. (1999). Training to improve virtual team communication, *Information Systems* **9**: 271-289.

Wathne, K., Roos, J. & von Krogh, G. (1999). Towards a theory of knowledge transfer in a cooperative context, *in* von Krogh, G. & Roos, J., *Managing Knowledge – Perspectives on cooperation and competition*: 55-81.

Weber, Y. (1996). Corporate cultural fit and performance in mergers and acquisitions, *Human Relations* **49**(9): 1181-1202.

Weiber, R. & Mühlhaus, D. (2010). *Strukturgleichungsmodellierung. Eine anwendungsorientierte Einführung in die Kausalanalyse mit Hilfe in AMOS, SmartPLS und SPSS*, Springer, Heidelberg.

Weissenberger-Eibl, M. A. & Spieth, P. (2006). Knowledge transfer: Affected by organisational culture?, *Proceedings of I-KNOW '06*, Graz, Austria.

Welsh, J. A. & White, J. F. (1981). A Small Business is not a Little Big Business, *Harvard Business Review* **59**(4): 18-27.

Westeren, K.-I. (2006). Knowledge as a factor to improve competitiveness for a firm in rural Norway, *Paper to be presented at the 46th Congress of the European Regional Science Association*, Volos, Greece.

Wong, K. Y. & Aspinwall, E. (2005). An empirical study of the important factors for knowledge-management adoption in the SME sector, *Journal of Knowledge Management* **9**(3): 64-82.

Wruck, K. H. & Jensen, M. C. (1994). Science, specific knowledge, and total quality management, *Journal of Accounting and Economics* **18**(3): 247-287.

Xenikou, A & Furnham, A. (1996). A correlational and factor analytic study of four questionnaire measures of organizational culture, *Human Relations* **49**(3): 349-371.

Zahra, S. A. & George, G. (2002). Absorptive capacity: A review, reconceptualization, and extension, *Academy of Management Review* **27**(2): 185-203.

Zahra, S. A., Ucbasaran, D. & Newey, L. R. (2009). Social knowledge and SMEs' innovative gains from internationalization, European Management Review **6**: 81-93.

Zhao, Z. J. & Anand, J. (2009). A multilevel perspective on knowledge transfer: evidence from the Chinese automotive industry, *Strategic Management Journal* **30**(9): 959-983.

MIX
Papier aus verantwortungsvollen Quellen
Paper from responsible sources
FSC® C105338

If you have any concerns about our products,
you can contact us on
ProductSafety@springernature.com

In case Publisher is established outside the EU,
the EU authorized representative is:
**Springer Nature Customer Service Center GmbH
Europaplatz 3, 69115 Heidelberg, Germany**

Printed by Libri Plureos GmbH
in Hamburg, Germany